CASS LIBRARY OF AFRICAN STUDIES

TRAVELS AND NARRATIVES
No. 68
Editorial Adviser: JOHN RALPH WILLIS
Department of History, University of California, Berkeley

NARRATIVE OF AN EXPEDITION
INTO THE
INTERIOR OF AFRICA

W0234771

AFRICAN TRAVELS AND NARRATIVES

NARRATIVE

OF

AN EXPEDITION

INTO THE

INTERIOR OF AFRICA

BY THE RIVER NIGER,

IN THE STEAM-VESSELS QUORRA AND ALBURKAH

IN 1832, 1833 AND 1834

BY

MACGREGOR LAIRD AND R. A. K. OLDFIELD

IN TWO VOLUMES

VOLUME I

Routledge
Taylor & Francis Group

LONDON AND NEW YORK

First published 1971 by Frank Cass and Company Limited

First Edition 1837

2 Park Square, Milton Park, Abingdon, Oxfordshire OX14 4RN
52 Vanderbilt Avenue, New York, NY 10017

*Routledge is an imprint of the Taylor & Francis Group, an
informa business*

First issued in paperback 2019

ISBN 13: 978-0-7146-1826-5 (hbk)
ISBN 13: 978-1-138-01100-7 (pbk)

Engraved by T. Jeavons ALM from a Sketch by W. W. Conye LADL.

THE QUORRA AGROUND BELOW THE JUNCTION OF THE SHARY AND NIGER.

NARRATIVE

OF

AN EXPEDITION

INTO THE

INTERIOR OF AFRICA,

BY THE RIVER NIGER,

IN THE STEAM-VESSELS QUORRA AND ALBURKAH, IN 1832, 1833, AND 1834.

BY

MACGREGOR LAIRD AND R. A. K. OLDFIELD,

SURVIVING OFFICERS OF THE EXPEDITION.

IN TWO VOLUMES.

VOL. I.

LONDON:

RICHARD BENTLEY, NEW BURLINGTON STREET,
Publisher in Ordinary to Her Majesty.
1837.

TO THE

MERCHANTS AND PHILANTHROPISTS
OF GREAT BRITAIN,

In the hope that the attempt recorded in these Volumes, to establish a Commercial Intercourse with Central Africa, *viâ* the River Niger, may open new fields of enterprise to the Mercantile world, and of usefulness to those who labour for the amelioration of uncivilized man, the following Narratives are respectfully inscribed by

THE AUTHORS.

PREFACE.

THE following pages detail an attempt to open a direct commercial intercourse with the inhabitants of Central Africa. It is well known that the attempt ended in a complete failure to make that intercourse a profitable one, and was attended with a melancholy loss of life. As far as proving the navigability of the Niger, and the ease and facility with which that mighty stream may be used for the purposes of commerce, it was successful in no ordinary degree, considering the novelty of the undertaking, the complicated nature of a steam-vessel, and the excessive mortality of the crews.

I also hope that it has in some measure dispelled the mystery which has so long enveloped the interior of that interesting country, and that it has proved that any man with common sense and common ability may ascend and descend the main artery of Africa (provided he escapes the effects of the climate) with perfect safety, in moderate sized vessels, from the sea to Boussa.

That my successors in the same field may avoid the errors that were committed, and profit by the experience acquired, is the principal reason why these Narratives are published.

The parties who, I hope, may, and trust will be principally interested, are those who look upon the opening of Central Africa to the enterprise and capital of British merchants as likely to create new and extensive markets for our manufactured goods, and fresh sources whence to draw our supplies; and those who, viewing mankind as one great family, consider it their duty to raise their fellow-creatures from their present degraded, denationalised, and demoralised state, nearer to Him in whose image they were created.

As I consider the happiness of my fellow-creatures ought to be the great end of all enterprise, and as the misery I witnessed made a great impression on my mind, I shall offer no apology for having, in the concluding chapter, stated freely and unreservedly my opinions and sentiments regarding Africa! — perhaps their chief value consists in their having been formed upon the spot.

If the publication of these Narratives advances the cause of African civilisation in the slightest degree, I shall consider the money, the time, and the life that has been lost, as in some degree compensated.

Mr. Oldfield having returned to the coast of Africa, his Journal has not had the advantage of his supervision while passing through the press.

MACGREGOR LAIRD.

London, 1st June, 1837.

CONTENTS

OF THE FIRST VOLUME.

CHAPTER I.

CHAPTER II.

CHAPTER III.

CHAPTER IV.

CHAPTER V.

CHAPTER IX.

CHAPTER X.

CHAPTER XI.

CHAPTER XII.

MR. OLDFIELD'S JOURNAL.

CHAPTER I.

CHAPTER II.

CHAPTER III.

CHAPTER IV.

CHAPTER V.

CHAPTER VI.

ILLUSTRATIONS.

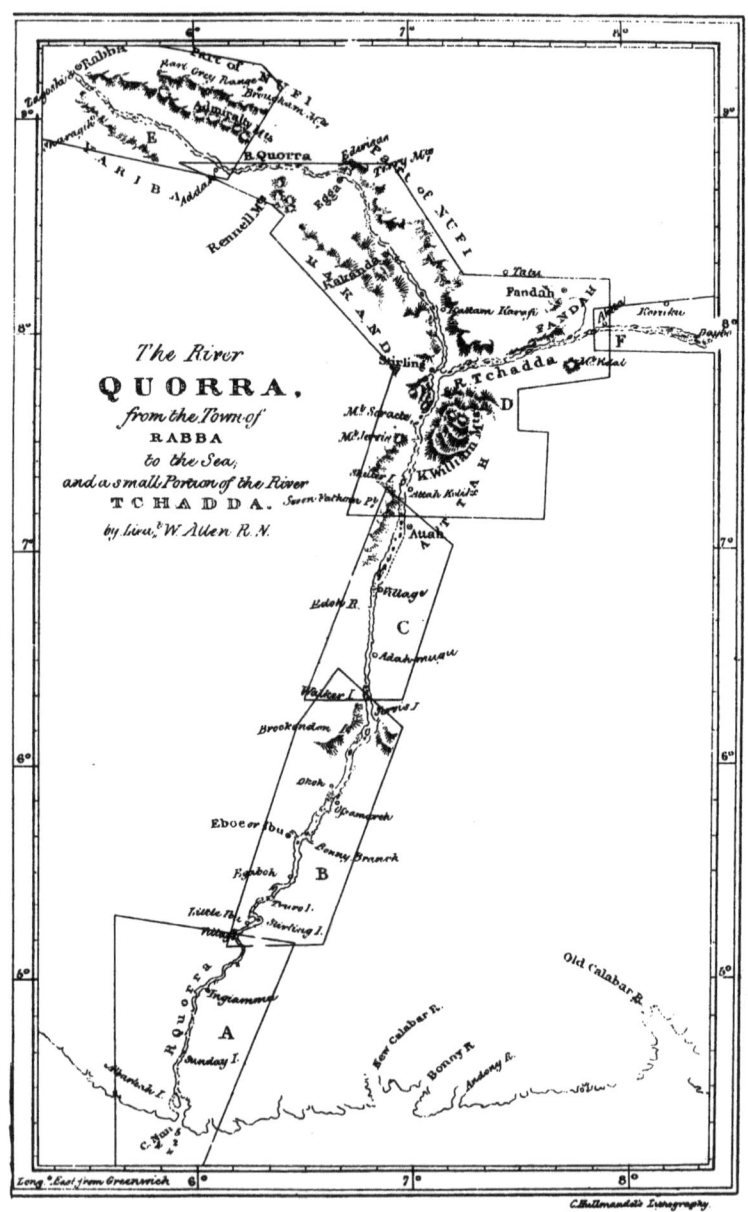

The River
QUORRA,
from the Town of
RABBA
to the Sea,
and a small Portion of the River
TCHADDA.
by Lieut. W. Allen R.N.

Long.ᵈ East from Greenwich

C.Hullmandel's Lithography.

London. Published by R. Bentley, 1837.

NARRATIVE.

CHAPTER I.

*Reasons for the Expedition.—The Company formed.—Mr.
Lander's Services engaged.—The Quorra and Alburkah
Steam-vessels and the Columbine Brig fitted and pro-
vided with Goods for the Country.—Plan of Proceeding.—
The Expedition joined by Lieutenant Allen and Mr. Jor-
dan.—Departure from Liverpool.—Stop at Milford.—En-
gineers of Steam-vessels considered.—Final Departure from
Milford.—Remarks on the Fitting of Steam-vessels for
Steaming and Sailing.—Madeira seen.—Anchor at Port
Praya.*

THE attention of British merchants has long
been turned to Central Africa as an untrod-
den, and therefore profitable field for commercial enterprise. The information obtained directly from English travellers, and indirectly
from captured slaves, describing that country as one on which Nature had profusely bestowed her choicest treasures, with the fact

that that immense territory, comprising one eighth part of the surface of the globe, had been for ages excluded from all direct communication with the civilised world, seemed to hold out the most flattering prospects of success to those fortunate individuals who should be the first to break into its secluded vales. The increase in the consumption of palm-oil in this country, the gradual decline in the supply of ivory, added to the inconvenience arising from the caprice and extortion of the petty chiefs at the mouths of the principal rivers of the African coast, stimulated the merchant to the discovery of new and unrestricted markets; while the knowledge that with the imperfect means of transit possessed by the natives, they already exported produce to the value of one million sterling annually to this country, presented to his mind the ready inference that a more direct intercourse between the producer and the consumer would tend to the benefit of both.

It will excite no surprise, then, that the splendid discovery of Lander was hailed with, if possible, more enthusiasm by mercantile than by scientific men. The long-sought-for highway into Central Africa was at length found, as open by the Niger as that by the Rhine, the Danube, the Mississippi, or the Oronooko, is into their respective countries. To the merchant it offered a boundless field for enterprise; to the manufacturer, an extensive market for his goods; and to the energy and ardour of youth, it presented the irresistible charms of novelty, danger, and adventure.

It must not, however, be supposed that these were the sole motives that actuated the promoters of that expedition, the proceedings of which are related in the following pages: other and nobler objects were connected with them. By introducing legitimate commerce with all its attendant blessings into the centre of the country, they knew that they were striking a mortal blow to that debasing and demoralising traffic which has for centuries cursed that unhappy land, and rendered some of the loveliest tracts on the face of the globe the habitation of wild beasts and noxious reptiles, or of man in a condition more disgusting and degraded than either. Under Providence, they aspired to become the means of rescuing millions of their fellow-men from the miseries of a religion characterised by violence and blood, by imparting

to them the truths of Christianity,—that mild
and beneficent faith, which proclaims " peace
on earth, and good-will towards man."

Urged by such considerations, the author, in
conjunction with several other gentlemen of Liverpool, determined to form a company, whose
first objects should be to open a direct communication with the interior of Africa; and, if this
were successful, to establish a permanent settlement at the junction of the Tchadda and
Niger, for the purpose of collecting the various
products of the country. In furtherance of
these views, a correspondence was opened with
Mr. Richard Lander, who had already signified
to one of the parties concerned his readiness
to embark in such an enterprise. An answer
was received from him, containing the strongest
possible confirmation, in his opinion, of their
most sanguine expectations. From his own personal observation, Mr. Lander assured the company that ivory, indigo, and other valuable
produce might be collected in any quantity at
a trifling expense.

The plan first proposed was to send out a
large vessel to the Nun river, fitted for the
palm-oil trade, with a small steam-vessel to

trade up the river. But after a personal communication with Mr. Lander, it was deemed advisable to depend solely on the trade in the interior for the returns; to fit out *two* steam-vessels of light draught of water, to carry it on in the upper part of the river, while the sailing vessel waited at the mouth to receive their cargoes.

As no steam-vessels, with the exception of a small iron one, were to be met with suitable for the purpose, it was determined to build a larger of the following dimensions :—length 112, beam 16, depth 8 feet, with a poop as high as the waist, to give the requisite accommodation for the officers. The keel of this vessel was laid on the 28th March, and by considerable exertions she was launched on the 29th May. Her engine was a single one of forty-horse power, manufactured by Messrs. Fawcett and Preston of Liverpool, who had a considerable share in the enterprise ; and it is but justice to these gentlemen to say, that the machinery of the engine, which commonly produces so much uneasiness to commanders of steam-vessels, was never for a moment throughout the voyage a subject of concern to any one : it seemed to

work as it were by instinct, and on leaving
Fernando Po, although I had lost my engineers
upwards of nine months before, it was still in
perfect order.

The smaller boat formed in herself the test
of an experiment of the most interesting kind.
With the exception of her decks, she was constructed entirely of wrought iron ; her bottom
was one fourth of an inch in thickness, and her
sides from three sixteenths to one eighth. She
was built at a considerable distance from the
water, and was conveyed on a truck through
the streets of Liverpool to the river. Her dimensions were, length 70, beam 13, depth $6\frac{1}{2}$
feet, with an engine of sixteen-horse power.

That such a vessel would be invaluable in
river navigation we were quite aware ; but whether she could withstand the wear and tear of
a sea voyage of four thousand miles, was at
best problematical. The result has added
another to many proofs that a theory fairly
grounded on scientific principles may safely
be brought into practice even against the advice and opinions of those who are generally
supposed to be the most able to give the one

or to form the other. Never were men more ridiculed than the gentlemen of the company when fitting out this vessel for her voyage to Africa. It was gravely asserted that the working in a sea-way would shake the rivets out of the iron of which she was composed: the heat of a tropical sun would bake alive her unhappy crew as if they were in an oven; and the first tornado she might encounter would hurl its lightnings upon a conductor evidently sent forth to brave its power. But what was the actual result? In spite of these wise opinions, her rivets are yet firm in their places, as the fact of her not having made a cupful of water sufficiently proves. Being in herself a universal conductor, she was always at the same temperature as the water in which she floated; and, for the same reason, though the lightning might play round her sides, it never could get on board of her.

The large boat was named the Quorra, the title given by the natives to a part of the river — or rather, I believe, a general term applied by them to all running waters. The smaller boat was called the Alburkah, a Houssa word signifying " blessing." The crew of the **Quorra**, including the officers, consisted of twenty-six men.

1 Captain.	2 Engineers.
2 Mates.	2 Firemen.
1 Purser.	2 Stewards.
1 Medical gentleman.	1 Cook.
1 Boatswain.	10 Seamen.
1 Carpenter.	2 Apprentices.

Total 26.

The crew of the Alburkah was formed of

1 Captain.	2 Firemen.
1 Mate.	1 Steward.
1 Surgeon.	1 Cook.
1 Engineer.	6 Seamen.

Total 14.

The Quorra was armed with one twenty-four-pound swivel gun forward, one eighteen-pound carronade, also mounted on a swivel, on the poop, and eight four-pound carriage guns on the main deck, with a proper proportion of small arms, not only for the crew, but also for the Kroomen, who, it was intended, should be shipped on the coast.

The Alburkah had one nine-pound swivel gun forward, and six one-pound swivels on her sides. Both vessels had their sides armed with chevaux-de-frise, for the double purpose of resisting boarders, and preventing the natives from crowding the decks ; a precaution which I strongly recommend to any vessels fitting out for a similar voyage.

The investments of goods were selected by Mr. Lander, who, it was presumed, from his acquaintance with the interior of Africa, was the most proper person for that purpose. No expense was spared in providing not only necessaries, but luxuries, for the officers and crew ; and it was generally allowed, that better-found vessels had never left the port of Liverpool.

The sailing vessel purchased by the company, named the Columbine, was a remarkably fine brig, of about two hundred tons, only four years old, and well provided with stores. She was appointed to carry out the goods, and was to remain at the mouth of the river to receive any cargo which the steam-vessels might bring down from the interior. She was to accompany the steam-vessels on the voyage out, as, having been built expressly for river navigation, it was not to be expected that they would prove good sea-boats.

In consequence of the cholera being very prevalent in Liverpool, our orders directed us

to proceed to Milford, which place at that time
was still uninfected : we were to remain there
till we had ascertained that we were perfectly
free from the disease, and then to proceed in
company to the river Nun, calling at Cape
Coast Castle for Kroomen and interpreters. In
case of parting company, the harbour of Port
Praya in the island of St. Jago was the appointed rendezvous.

Before leaving Liverpool we were joined by
Lieutenant William Allen, R.N. a gentleman for
whom the admiralty had requested a passage in
one of the vessels, for the purpose of making a
survey of the river. The company had very
liberally granted this request, although they had
not received the slightest assistance or encou-
ragement from any department of the government. The admiralty remained pledged on the
occasion that no information should be given
to the public by this officer without the express
permission of the company. About the same
time we were also joined by Mr. Jordan, a
young man of amiable manners and considerable
acquirements, whom it was intended to establish
in the interior of the country, provided his health
could withstand the effects of the climate; but

owing to there not being sufficient room in the
Quorra's cabin, he made the voyage out in the
Columbine. Mr. Lander passed several weeks
in Liverpool while the vessels were preparing,
but left them a short time before we sailed,
agreeing to join the expedition, of which he was
to take the command, at Milford.

In such a purely commercial place as Liverpool, any enterprise that may lead to an extension of its commerce is viewed with far more
interest than persons unconnected with business
can well imagine. The speculator regrets that
he has no share in the anticipated success,—the
cool and wary merchant calculates the time that
may elapse before he can avail himself of any
discoveries that may be made,—the young and
enthusiastic envy the share of those of their own
age who, released from the thraldom of the
counting-house, are about to enter upon a new
field of enterprise and honourable danger,—while
the old and experienced, if they doubt its success,
cannot bring themselves to damp by gloomy
forebodings the buoyant and adventurous spirit
of those engaged in it. In addition then to
that of private friends and connexions, we had
also the sympathy and good-will of the public;

and I will not attempt to conceal that in my own case, at least, the consciousness of this alleviated in some degree the regret I felt on leaving those who were nearly and dearly allied to me.

We departed from Liverpool on Wednesday the 19th July 1832, in company with the Alburkah and Columbine, the latter being in tow of a steam-vessel belonging to the City of Dublin Company. A fresh breeze had set in from N.N.W. and I was much pleased to find that my expectations respecting the liveliness of the Quorra and Alburkah, and particularly the latter, were more than realised, neither of them shipping any water though steaming end on to a short tidal sea. Before the pilot left us, Captain Harries mustered all hands on deck; and I was particularly struck by their appearance of perfect sobriety. To the reader this may appear nothing more than regular, but it is a most uncommon thing in a merchant ship leaving port. The crew were all picked men from twenty-five to thirty-five years of age; and little did I think, as I beheld their athletic and powerful frames, that in a few months the only survivors of us all would be myself and three others.

After a very good passage of thirty-six hours

we arrived at Milford. A few hours later, the Alburkah came into the harbour; and her commander Captain Hill fully corroborated by his testimony my good opinion of his vessel. The Columbine not being able to get round Holy Head the day she sailed, was forty-eight hours longer on the passage.

During our stay at Milford, where we were detained until the 29th, waiting the arrival of Mr. Lander, we discharged one of the engineers, and entered another from one of the post-office packets. The great demand for this description of men, and the high rate of wages they can get on shore, render it very difficult for vessels engaged on foreign service to obtain them: we were obliged to pay 16*l.* per month for men who on shore would not have earned more than 30*s.* per week. This is a great drawback on the introduction of steam-vessels generally abroad; and until the profession of mechanical engineering is considered a fit pursuit for respectable young men, it must remain so. On board a steam-vessel every officer should be perfectly independent of the engineer; each should be capable of working the engines, and of judging whether the different parts of the machinery are kept in proper order or not. The importance of having some public means of instruction in this branch of practical science must be manifest to every one. The general adoption of steam power in the foreign packet service, and the large capital invested in private steam-vessels, hold out sufficient inducement to individuals who make the profession their study; and it only requires government to make it an honourable pursuit, by establishing a class of master engineers, to draw hundreds of young gentlemen into it who are now frittering away their time waiting for commissions in the army or navy. Surely there can be no great difference between adjusting a screw or packing a piston, and examining a splice or scrutinising a knapsack.

Our stay at Milford was rendered agreeable by the hospitality and kindness of Mr. Leach and Captain Chappell. The latter gentleman, as agent for his Majesty's post-office packets, granted us the use of their hulk, which enabled Captain Harries to get our decks cleared and stowed for sea, and to fill up with coals to the very moment of our departure. To both these gentlemen my grateful thanks are due for their kind attentions, which were the more unexpected as we were personally unknown to them.

On the 25th, the wind being favourable and
there being no appearance of Mr. Lander, nor
any letters from him, we despatched the Columbine and Alburkah, towing them both out of the
harbour. Their orders directed them to make the
best of their way to Port Praya, and there to wait
the arrival of the Quorra.

On the evening of the 29th we were agreeably
surprised by the appearance of Mr. Lander, and
in less than an hour afterwards were under weigh
with steam and sails, a light breeze having set
in from the N.E. We continued under steam
till we lost soundings; we then disconnected the
engine, and with the freshening breeze stood
across the bay.

The difference between a sailing vessel and
a steamer was surely never more perceptible.
From going merrily along at the rate of eight
or nine knots per hour, with the paddle, as if
instinct with life, dashing and flapping the blue
waves of the ocean, the speed of our vessel suddenly slackened into that of a quiet *jog-trot*, and
she went on rolling lazily over the long swell at
the rate of not more than two or three. Having
had the planning and fitting out of the vessel,
I was perhaps more annoyed than any of my

companions at witnessing her performance under
canvass ; and as I knew the fault could not be in
her shape, as she was remarkably fast for her
power under steam, I could attribute it to nothing but the want of proper masts and spars.

In all voyages where steam is to be depended
on for the whole distance, the lighter the masts
and rigging are the better. In fact, a steam-
vessel then merely requires masts sufficiently
lofty to enable her to set her fore and aft sails
so as to steady her when she has the sea on her
beam. But in a voyage where the dependence
of performing it is partly on steam and partly on
canvass, a totally different rig is required : the
masts should be loftier in proportion to the beam
than even in a sailing vessel, and the lower yards
should be proportionably square. The Quorra
unfortunately was rigged in the Channel fashion,
and we were obliged to resort to the most laughable devices to get her along. In vain were
spare sails set on the funnel, and awnings run out
as lower steering sails : under the most we could
do for her, and the most favourable circumstances,
we never could get even five miles per hour out
of her under sail. I have never seen it tried, but
I should think that a square mainsail set flying

would be of great use to a steam-vessel on a long voyage.

The inconvenience attending the method we had adopted in rigging the Quorra, was increased by the paddle-boxes being so near the surface of the water as to check her way materially at every roll the vessel made. To remedy this in some measure, the after ends of them were cut away, to allow a free passage for the water; and this produced a beneficial effect.

The management of the paddle when disconnected from the engine was the subject of several experiments. The general practice is to remove the floats and expose merely the iron rim and arms of the wheel to the action of the water; but we found it better to allow the floats to remain on, as a much less velocity then overcame the friction in the gudgeons than by the former plan, and the vessel's steering was also considerably improved by it. The results, as nearly as we could ascertain, were as follows:—Without the floats, the paddle-wheels revolved when the vessel had a velocity of four and a half miles per hour; with the floats fixed, they required only a velocity of one and a half mile per hour to set them in motion; the vessel steered much easier, there

was less strain upon the paddle-boxes, and if
the paddles were to be affixed to the engine, it
was done without the delay occasioned by having
the floats to attach.

I have been more particular in stating this,
as I have known a great deal of money thrown
away in sending out steam-vessels to foreign stations with their paddle-boxes built up and their
paddles stowed on board. The safest and the
simplest plan appears to be, to complete them
as steam-vessels in this country, and after they
are fairly clear of the land, to disconnect the paddles from the engine and allow them free play in
their gudgeons. They will stop the vessel's way
very little, and there is the advantage of always
having the engine to fall back upon in case of
being becalmed or getting too near the land.
We had, indeed, a striking proof of this on our
voyage out. In running down the N. E. trade
we found ourselves becalmed under the island of
Palma, though it was eight or ten miles distant
from us : there was a fine breeze about two miles
a-head. The steam was got up, the paddles
were connected, and in ten minutes after the engine was at work we had as much wind as the
vessel would stand. The steam was then let

down, the boilers were blown off, the paddles were disconnected, and we were under sail again in less than two hours from the time we had been becalmed.

Fortunately we had fine weather in crossing the bay. In twelve days after we had left Milford we were in sight of Madeira, having had a fair wind the whole way. This beautiful island, with its bold and lofty outline, assumed a thousand different forms as, impelled by the first faint breathing of the N. E. trade wind, we sailed slowly by it, and at every changing prospect felt the full justice of the encomiums we had heard lavished on its charms. Indeed, so attractive did this isolated little paradise appear to us, even from the distance at which we viewed it, that we fain would have lingered on our way to enjoy the contemplation of its beauties. The balmy freshness of the air—the serenity of the sky—the tranquil and engaging appearance of the island, reposing as it were on the bosom of the ocean as the abode of peace and happiness, produced in our minds sensations of delight, and formed a grateful contrast to the bustle and turmoil—the inconvenience and confinement of a ship. On the 17th we anchored in Port Praya.

CHAPTER II.

AT the time we visited Port Praya, the Cape
Verd Islands were suffering from the effects of a
grievous famine, arising from the want of rain,
none having fallen in any considerable quantity,
as we were informed, for three years.   One thousand people had died in St. Jago, and upwards
of fifteen hundred in Fogo, a smaller island, not
so much frequented by ships, and consequently
more dependant on its own resources.   The governor was in politics one of Don Miguel's party,

and, for political reasons, refused to receive the visits of any Englishman!

I never beheld such an assemblage of wretched and emaciated forms as crowded the streets and beach of Port Praya on our arrival. Regretting our own inability to supply the wants of these miserable objects, we could not abstain from reflecting in no measured terms on the conduct of a government, which could allow its colonists to be dependant for their very existence on the casual visits of strangers for food, while those very colonists were within a week's sail of the African coast. Captain Harries, who had been here five years before with the African squadron, described the place as then presenting the appearance of a garden; and the natives assured us that forty-eight hours' rain would suffice to change the face of the country. I am sure it stood much in need of it;—all signs of vegetation had disappeared, and everything was scorched and withered.

This was my first introduction to a slave population, and I was less struck by the appearance of the slaves than I had expected, being fresh from a land where nakedness is never exposed save to excite pity or disgust : but a black man in a state of nature does not particularly strike one with any idea of indelicacy; and he will walk deliberately by you in all the dignity of nature, his limbs unencumbered with dress, proudly yet unconsciously asserting his right to be considered one of the lords of creation. I could not help thinking, whenever I saw the men strutting about in this manner, that a white man would cut but a sorry figure under similar circumstances. It did not excite much sympathy in me to see the men driven to their work and obliged to perform it, for they evidently did not get through one fourth part of that which a labourer would do in England; but it was disgusting to the feelings to see the women toiling up the steep ascent which leads to the town, with heavy loads of salt upon their heads, many of them being in a state calculated to excite the sympathy of any one not entirely devoid of feeling.

There is a capital well behind the town, and thither Dr. Briggs and I repaired for the purpose of seeing the domestic slaves. They certainly appeared to be a very happy and contented race. I had often heard of African tongues, but no description could do justice to the confused clattering noise which these creatures made. We sat contemplating them for some time, and were much amused by their half Portuguese, half English sallies at our expense.

The harbour of Port Praya abounds in fish; but although the people here were absolutely starving, not more than one or two boats were employed in fishing. I learned, however, that the unfortunate natives, being all slaves, were not allowed to keep canoes or boats of any kind, lest they should take it into their heads to attempt their escape to the African coast. If this were really true, then had their owners to answer for the double guilt of enslaving and starving them; and in the present instance they most deservedly shared the evil arising from such undue precaution, for disease and death were making great havoc among the Portuguese as well as the slave population, and both were alike unable to avail themselves of that relief which Providence had placed within their reach.

A small American schooner was lying in the bay at the time we were there, and I could not help admiring the enterprise and spirit displayed by Jonathan in venturing across the Atlantic in such a craft. She had a cargo of what he called " notions;" which notions consisted of a deck-load of timber, staves, hoops, and baskets, strung up to the cross-trees ; tobacco, flour, household furniture, ironmongery, crockery, old clothes, salt fish, &c. &c. He was bartering these for goat-skins and other produce of the island : he had disposed of the chief part of them, and, as he expressed it, was going "slick home" for a cargo of Indian corn, which he calculated "would pay considerable if the starvation progressed." He sailed his schooner on temperance principles, which, as his crew consisted of only a mate and two boys, showed at once his good sense and right feeling.

Having received our water on board, we were under weigh, in company with the Alburkah and Columbine, on the evening of the 19th of August, glad to leave a place where we had only witnessed misery without being able to alleviate it. We continued under steam for some hours, and then, the wind being favourable, we disconnected the engine from the paddles and made the best of the way we could under canvass, the brig running with us under her topsails with the yards on the cap. On the 20th the wind fell, and from that time until the 1st September we had a succession of calms, with a heavy swell from the S. W.

accompanied by the heaviest rain I ever witnessed. Captain Harries took the brig in tow
several times, but the swell strained the Quorra
severely. Under these circumstances, finding
that we were being set by the swell and current
considerably to leeward, Captain Harries determined to bear up for the Isles de Los, to obtain a
supply of coals there from the Columbine, and
attempt to weather the shoals of St. Ann under
steam, after which he expected to be able to hold
his way down the coast under canvass. Under
the circumstances of our situation, I am not
aware that he could have acted otherwise—the
Quorra under canvass being worse than useless
unless the wind was abaft the beam.

We arrived at the Isles de Los on Sunday the
2nd September, and anchored between Factory
Island and the mainland. The Alburkah did not
join us till the afternoon, having being driven
still further to leeward. I was much struck with
the extreme beauty and romantic appearance of
these islands : the rich verdure with which they
were covered even to the water's edge contrasted
forcibly with the brown and cinder-like appearance of those we had just left. There were no
white people on the islands at the time of our

arrival, the British settlement there having been long previously broken up. The inhabitants whom we found there consisted of discharged veterans from the West India regiments, and the colonial schooner from Sierra Leone was lying at Tamara, the principal island, paying them their pensions. The day after our arrival, his Majesty's brig Charybdis, Lieut. Crawford commander, came in on her way to the Gambia, and anchored near us. Lieut. Crawford informed us that the slave-trade was brisker than ever on the coast. He had been unfortunate, not having taken any prizes, and appeared to regret very much having to go to the northward, where there was less chance of falling in with them.

The Isles de Los are well situated for trade in consequence of their proximity to the mouths of the Rio Pongo and Rio Nunez, two rivers hitherto little frequented by the English, though a considerable trade has always been carried on with them by the Spaniards and Portuguese. The islands are also considerably healthier than any part of the neighbouring coast; they are evidently volcanic, and appeared to me to have formed the sides of an immense crater.

On Friday the 7th of September, having com-

pleted our coals and water, we departed from
these islands, having previously arranged with
the Columbine and Alburkah, in case of sepa-
rating during the night, that Cape Coast Castle
should be our rendezvous. In the evening the
weather became bad and squally, and we exchanged blue lights with our consorts. Next
morning, the brig not being in sight, we concluded that she had stood out to sea to weather the
shoals of St. Ann, which afterwards proved to be
the case. The Alburkah having fallen considerably to leeward, we took her in tow for some time;
when our fuel being nearly exhausted, we came to
an anchor in fifteen fathoms water, let down the
steam and disconnected the paddles, hoping that
we might get off the shore with the land-wind.
At midnight we started again with a light breeze,
and at six A. M. found that the current had set us
considerably to leeward, and the Alburkah still
more so. The latter vessel's bottom had become
so foul that her speed was considerably dete-
riorated by it : this circumstance, although it
proved that oxidation was not going on, was annoying, as it delayed us, and a little foresight
might have prevented it.

Finding it impossible to get the vessels to their

destination under canvass, Captain Harries determined to anchor under York (a branch settlement of Sierra Leone), then under our lee, and,
if possible, to get Kroomen there to supply the
vessels with wood on our voyage to Cape Coast
Castle, which it was too probable would be a
coasting one. On Sunday the 9th, we anchored
within a mile of the shore in company with the
Alburkah. York is one of those villages dependant on Sierra Leone, where the liberated Africans
and discharged soldiers from the West India regiments are located. It is beautifully situated on
a hill, the superintendant's house and the school-
room crowning the summit. Sunday being the
day of our arrival, we found the inhabitants
dressed in their best attire, which gave them, and
more especially the women, a respectable and prepossessing appearance. We immediately called
on Mr. Pratt, the superintendant, and were much
disappointed to find that he was from home, as
we could make no arrangements for procuring
either wood or Kroomen until his return.

In his absence we were welcomed by his house-
keeper, who with true colonial hospitality pressed us to remain on shore, saying he might possibly arrive during the night. Mr. Lander, Lieu-

tenant Allen, and myself, accordingly took quiet possession of his house, and were making ourselves exceedingly comfortable, when about nine o'clock we were surprised by the sudden appearance of Captain Harries, who, not being aware of our intention of staying all night, had brought a boat on shore for us. Immediately on landing he was seized by the town-guard, who, it seemed, had been watching our movements with some suspicion during the day, and now appeared in considerable force to insist upon our returning on board. A most laughable scene ensued, which ended in our being turned out of our comfortable quarters and marched under escort down to our boat,—our fair hostess alternately abusing our conductors for their breach of hospitality, and threatening them with the consequence of their conduct.

Their reasons, however, appeared very good when they were afterwards explained to us. It is not uncommon for the liberated Africans to be kidnapped by slave-vessels even when under British protection, and our appearance being, to say the least, novel, if not suspicious, the inhabitants in the absence of Mr. Pratt, beat to arms for their own protection.

In the morning Captain Harries landed a party
on a small island to cut wood, and we went on
shore to see the town and neighbourhood. The
view from the superintendant's house is grand
and imposing. To the west and south lay the
sea, with the Banana Islands and the entrance
to the Sherborough river; and behind the house
were mountains rising almost perpendicularly
from their base, clothed with forests of immense
trees to their very summits. A large tract of
cleared and partially cultivated ground extends
from the base of the hill, on which the government house is situated, to the banks of a mountain stream which forms the boundary of the
settlement. The land did not appear to be very
good, being too much washed by the mountain
torrents during the rainy season. Ginger
seemed to be the principal plant cultivated for
export, and of that but a small quantity was in
the ground.

The luxuriance of tropical vegetation has a
most imposing effect on a person the first time
he beholds it; while the thought that that very
luxuriance, the absence of which is deplored in
many countries, forms the bane of this, forcibly
reminds him of the impotence of human reason

to fathom the designs of Omnipotence. The wild cotton and oak trees grow to an immense size, and are so intermixed with parasite plants as to form an impervious shade. The stillness which reigns throughout the scene is another of its characteristic features,—one which chills the heart, and imparts a feeling of loneliness which can be shaken off only by a strong effort.

After a long walk we returned to the village, and were welcomed with much kindness and cordiality by Mr. Pratt, who had arrived during our absence. He informed us that he had no Kroomen, and would have to send to Free Town for them, to which place, it being only thirty miles distant, Captain Harries determined at once to take the Quorra and select the men himself. We were also in want of provisions for such an addition to our number, as it seemed probable that the voyage might be much more protracted than we had at first anticipated, and we were uncertain where we might fall in with the brig, which had the bulk of our stores on board.

Mr. Pratt's return having restored the confidence of the people, we passed a pleasant evening on shore, Lieut. Allen, Mr. Lander, and myself remaining all night. The two former

gentlemen were seized during the night with
spasmodic colic; which so alarmed Mr. Pratt
and me, that we despatched a canoe for Dr.
Briggs of the Quorra, who came on shore immediately in a torrent of rain, and by administering
powerful remedies in some measure relieved
them, though both were much weakened by the
violence of the attack.

On the 12th, having taken sufficient fuel on
board, we ran up to Free Town, leaving the
Alburkah at York. Here we were immediately
boarded by some of the gentlemen belonging to
the colony, who had heard of our being at the
Isles de Los from the schooner we had left there.

No sooner had we anchored off Free Town
than our decks were crowded with Kroomen anxiously beseeching us for employment. Among
others were some bumboat women screaming out
recommendations of their fruit and palm-wine,
and washerwomen of rather equivocal appearance
proffering their services. Captain Harries immediately selected ten Kroomen who had previously served on board his Majesty's vessels, and
had good characters from their old commanders.
Before engaging them, they were informed of
the nature of the service for which they were

required, and I was much struck by the simplicity and manliness of their answers. " Englishman go to debil, Krooman go with him," was their general reply ; a proof not only of the inherent bravery of the people, but of their confidence in the character of an Englishman, and, let me add, also a tacit admonition to their employers of the responsibility they are under to treat them well, and, by setting them a proper example, to preserve their good opinion. As I consider myself under great obligations to this class of men, I may be excused for adding a short account of their country, their habits, and pursuits.

The Kroo country extends from Simon River along the coast to Cape Palmas, and from thence to Cape Lahoo. The inhabitants of that district consist of two distinct classes, namely, Kroomen and Fishmen ; the former being the best axemen and servants on shore, and the latter excelling as boatmen and sailors. The dexterity of the Fishmen in the water is quite astonishing. They avail themselves of their superior skill in this respect to waylay the Kroomen on their return from Sierra Leone in the small canoes in which they are accustomed to make the passage along the coast laden with

goods, the produce of their wages. The Fishmen exact a tribute from the Kroomen when
passing their shores; and if their demands be
not complied with, they will upset their canoes,
and from their superior agility in the water
generally manage to secure the greater portion
of the cargoes. In Sierra Leone they inhabit a
small village close to Free Town, and keep themselves apart from the emancipated negroes, on
whom they look down with most sovereign contempt. Their mode of life is very peculiar.
Their own country producing barely sufficient to
support them, every Krooman or Fishman leaves
home at the age of thirteen or fourteen, under
the care and patronage of a headman, who con-
ducts him to Sierra Leone and takes him on
board of any ship in which he may happen to be
employed on the coast, with the rest of the boys
who may have been placed under his management, and who generally amount to eight or ten
in number. The headman receives their wages,
keeps them in order, flogs them when required
to do so, and after a certain period they are at
liberty to work on their own account. It is in
fact a regular system of apprenticeship.

A collection of these headmen reside at Free

Town; and every ship, immediately she arrives, is boarded by them with their boys clamouring for employment. The agreement is made with the headman, who selects the number of boys required, and trots them out for inspection; and as they are not much encumbered with clothing, a single glance is sufficient to discover those that will suit. Three yards of cloth suffice for the equipment of each, and they are ready to go to sea immediately. The wages at Sierra Leone are four dollars per month paid in goods, or about two and a half dollars sterling. These people have the common failing of negroes—they are given to lying and pilfering; but they never desert their employers in danger or distress; they are constitutionally brave, and are easily kept in order: they are the life and soul of the trade on the coast; without them the cargoes could not be stowed, nor could boats be manned.

The Kroomen are the principal woodcutters in Sierra Leone, and great numbers are constantly employed in the squaring and floating of African teak to the ships. This timber is found in great abundance on the banks of the river of Sierra Leone and its tributaries, and is in fact the only production of the colony that forms an

article of considerable exportation. Kroomen
are also employed in great numbers as trading
men on the coast; and it is said that where
confidence is reposed in them, they are found
not undeserving of it. Their moral character is
better than that of their brethren the Fishmen;
but they are not so fine a race of men, and are
comparatively useless on board a ship.

The great object of the Krooman or the Fishman is to get as many wives as will support him
in idleness in his own country. The instant,
therefore, that he has earned and stolen money
enough to purchase one, he sets out with two
or three of his companions in a canoe for the
Kroo country. On his arrival he lays his treasures before the father of his intended; and if
they are considered sufficient, he enjoys her
society for a week. He then leaves her with
her family to set about obtaining another by his
next trip; he returns to Sierra Leone, and goes
on board a ship upon the coast to earn the price
of a second wife; and he repeats the process
until he has a colony of wives at Cape Palmas,
or in its neighbourhood, and then retires from
active life, not upon his fortune, but upon his

wives, they supporting him in dignified idleness by their labour.

The difference between the negroes generally and the Kroomen appears to be, that the latter may be stimulated to immediate exertion for the sake of future gain, while the negroes can be excited only by existing necessity or fear. The appearance of the Kroomen is much superior to that of any other race of negroes on the coast; they are generally tall and well-proportioned, their limbs are muscular, their gait erect and firm. I was informed that they are never taken as slaves, in consequence of the unyielding spirit they have always displayed; and indeed there is a certain air about them which proves that they were born free and will still remain so. On board ship they are punished when necessary by their headman, who flogs them most severely on its being required by his captain; but it is considered dangerous to interfere with his office in this particular, as instances have been known of their jumping overboard after being flogged by a white man. Considering them altogether, they are decidedly the most interesting race of people to be found on the African coast, and if they were pro-

perly educated, might be made exceedingly useful
as agents to forward the great design of civilising
Africa.

In the morning of the 13th of September, Captain Harries waited on the governor, Colonel
Findlay, who politely ordered that we should be
supplied with provisions from the government
stores.   At three o'clock we left Free Town, accompanied a short distance by several gentlemen
belonging to the colony, and in the evening arrived at York.

On the 15th, having bidden farewell to our
kind and hospitable host, Mr. Pratt, we departed
from York, taking the Alburkah in tow ; and passing the Banana Islands, steered across the shoals
of St. Ann, through a narrow and rather intricate
channel, an outline sketch of which had been
given to Captain Harries by Mr. M'Cormack of
Sierra Leone.   These shoals are the only dangers of any importance between Sierra Leone
and the equator, and from the shifting nature
of their sands are always allowed, as seamen express it, " a wide berth."

After steaming thirty-six hours, we disconnected the engine and made sail, expecting from
the course which the vessel lay to have a quiet

passage to Cape Coast. On Tuesday the 19th we made Cape Mesurado, having fallen about forty-five miles to leeward while under canvass. On our arrival in the bay off Monrovia, the American settlement, we found two American brigs and a schooner lying at anchor, the latter belonging to the colony.

The appearance of Monrovia from the sea is picturesque. Cape Mesurado, on the side of which it stands, rises abruptly from a low flat shore, and towards the sea is almost perpendicular. As our only object was to obtain a supply of fuel, we were much disappointed on finding that the river St. Paul was not considered navigable, no vessel, with the exception of a small schooner of fifty tons burthen, having crossed the bar. As it was impossible to get wood from off the beach without great delay, Captain Harries determined to attempt the bar; and accordingly having sounded it, and finding seven feet water, we crossed in safety and found ourselves in a small shallow river full of rocky shoals. On the arrival of the Alburkah we proceeded about two miles and a half up the river, to Caldwell, another settlement of the Americans. The governor had sent a gentleman of colour with us to obtain supplies of wood; and immediately on his arrival, the inhabitants went to work and cut sufficient wood in two days to fill both vessels. On the 24th we recrossed the bar, and proceeded on our voyage with the Alburkah in tow.

As Liberia is a settlement that has been much talked of, having excited a great deal of interest in this country, I am induced to record the impression it made on my mind during our short stay there. It is well known that the object of this colony is professedly the civilisation of Africa, by the introduction of free American negroes as voluntary settlers, under a democratical form of government. But let us inquire how this object is carried into execution. In the American States, particularly those of the South, the life of a free negro is embittered by every means that the ingenuity of a slave-holder can devise: he is constantly liable to insult and oppression, and is always looked on with suspicion and distrust;— suspicion, from the consciousness of his oppressor that he has much to avenge,—distrust, from the increase of their number, wealth, and intelligence. These appeared to me the principal motives for establishing the colony of Liberia. The free negroes of the Southern States of America were becoming too numerous, and therefore dangerous, and Liberia in Africa was colonised to perpetuate the existence of slavery in the Western hemisphere. It may be said that free negroes are not forced to go to Liberia, and perhaps actual force may not have been, and may not be resorted to;—but what is the case?—first, their lives are rendered miserable by a continued series of petty oppressions, and then Liberia is held out to them as a sort of paradise of plenty and freedom where no white man is to be allowed to tyrannise or oppress. An intelligent mulatto said to me on my questioning him on the subject, "It was not exactly kidnapping, but we were inveigled away under false pretences."

As to civilising Africa by means of Liberia, it is well known that from the time the colony was first established it was constantly at war with the natives, until their partial extermination left the strangers in peaceable possession. It is true they assert that they bought the ground with the right of extension; but this only adds hypocrisy to cruelty.

As to the commercial situation of the colony, the Americans have certainly not shown their usual acuteness in the choice of natural advan-

tages : instead of fixing their head-quarters at
the mouth of some considerable river, they have
taken possession of the most unhealthy and sterile
part of the coast, with no inlet into the interior;
the consequence of which is, that they are dependent on the uncivilised negroes on the Grain
Coast for their supplies of food.   It is certainly
not to be expected that an infant colony can support itself for the first few years ; but it appears
questionable whether Liberia will ever raise food
sufficient for a very moderate population, and it
certainly never can export any quantity of tropical produce.   During the time we remained
in the river St. Paul our vessels were crowded
by respectable and intelligent mulattoes, all of
whom, with the exception of the coloured editor
of the Liberia Gazette, and one or two others
in the pay of the Society by whom they are sent
from America, complained bitterly of the deceit
that had been practised towards them, and of the
privations under which they were then suffering.
It was often a source of regret to me during our
voyage that I had not acceded to their wishes,
and taken some of them on board our vessels, as
they were fine intelligent men who would have

been invaluable to us in the interior; and I strongly recommend, in any future expeditions of this kind, that the crews be completed here in preference to taking Europeans as sailors and engineers.

It is needless to say, that the foregoing remarks are not meant to reflect on the motives of those gentlemen both in America and England who advocate the cause of a free negro settlement. No doubt they acted on the information they received; but I am not aware that any of them have personally witnessed the actual working of the system. The principle of negro colonisation is admirable; and if the money and life that have been expended on Liberia had been properly applied, the results would, ere now, have been very different. I have no hesitation in affirming, that if the Americans had formed Liberia either at the mouth of the Rio Ponga, the Rio Nunez, or the Rio Grande on the windward coast, or on Fernando Po, Corisco, or Cameroons on the leeward coast, it would have been a thriving and independent colony by this time: there never was a collection of men better formed and adapted by nature and circumstances for African colonists

than the American negroes—they have all the
acuteness and enterprise of the Yankee grafted
on an African constitution.

On the 27th, our fuel being expended, we
were again under the necessity of entering one
of the small rivers which are so numerous on the
coast for a fresh supply : it was called by the natives Simon River.   We found the banks inhabited by a mixed race of Kroomen and negroes.
It is a noted slaving place, and the natives had
great quantities of firewood piled up in readiness
for a large slave-brig that was expected.   The
mode of collecting slaves on this part of the coast,
where one depôt is not sufficiently large to load
a vessel, differs from that on the leeward coast.
A slaver comes on the coast to windward about
the Gallinas River, and runs down as far as Cape
Palmas, calling at all the small slaving rivers
and ports, and leaving an assortment of trading
goods to purchase slaves, having arranged the
time the cargo is to be ready for shipment, which
is generally in five or six weeks.   The vessel then
runs off the coast, and, if boarded by any of our
cruisers, reports herself as bound for Prince's
or St. Thomas's Islands.   At the expiration of
the time she returns, and commencing at the

most windward depôt, in a few days completes her living cargo.

As soon as we had arrived in the river, we were visited by the chief of the eastern bank, who assured us that he only suffered his opposite neighbour, the chief of the western bank, to exist for compassion's sake. The only difference we could perceive in these two rival chiefs was, that one painted his face white, while the other painted his red ; but we were so far fortunate in having two vessels, as by assigning the supplies of the Alburkah for one, and those of the Quorra for the other, we were enabled to please both. By bartering cloth for firewood, we succeeded in filling both vessels by the 30th, as the natives are expert axemen, and had besides a large stock ready cut. Dr. Briggs and I availed ourselves of the leisure in this interval to take a long ramble in the country, and visited several of the huts : those we entered had a floor formed of split bamboos raised about two feet from the ground; they were clean and neat in their interior arrangements. The shores about this part are rocky and bold, and evidently of volcanic origin.

On Monday the 1st of October, we anchored

off Cape Palmas to complete our complement of
Kroomen, which we effected by taking ten fine
fellows in addition to those we had obtained at
Sierra Leone. Being anxious to see the head-
quarters of this interesting race, I tried to land;
but the boat I was in was nearly swamped by a
roller in a vain attempt to cross the bar of the
small river under the cape, and I was glad to
get on board the Quorra again into safety. I regretted not being able to land, as I had understood that many European articles were to be
found in the houses of the chief men, with which
they are regularly supplied by their countrymen
from Sierra Leone.

On the 6th of October we came to an anchor
off the Dutch fort of Axim. Having finished
our last stick of firewood, Mr. Lander and Lieut.
Allen went on board the Alburkah, she having
still some left, the former being anxious to reach
Cape Coast for the purpose of arranging matters
concerning his men. In the morning of the 7th
we went on shore, and were welcomed by the
governor of the fort, who was the only European
resident. His garrison consisted of one corporal
and five men, who, he told us, gave him infinitely
more concern and trouble than all the rest of his

subjects. We were not a little amused at the
method adopted by this son of Mars in administering his authority, which extends from Cape
Apollonia on the west to Dix Cove on the east.
His plan was simply as follows :—On receiving a
complaint from any one within his district, he immediately sends his stick to the village in which
the offender resides.   The messenger who is
charged with it places it upright in the ground
in the centre of the village, and remains close by
it.   The natives know well the meaning of this,
and that the appearance of the governor's stick in
this formal manner is nothing more or less than
a demand of eight ackeys, or half an ounce of
gold, from the offending person.   The messenger
remains by his charge for twenty-four hours if
the offender be obdurate ; at the expiration of
which time the governor's hat is despatched after
the stick, and is deliberately placed on it as it
stands in the ground.   This, however, makes the
matter more serious, and the demand is increased
to an ounce of gold in consequence.   Should the
hat and the stick be insufficient to move the offender to pay the fine, a third messenger appears
with the governor's sword, and an additional
ounce of gold is required ; and it is a remarkable

fact, that this last resource has never yet been
known to fail—in fact, the stick alone is generally all-sufficient.   In addition to this, there
is a certain charge for putting a prisoner into
irons, and another for taking him out again;
another for his lodging in the castle while confined, and another to the constable for locking up.
I had often heard of Dutch colonial tyranny,
but could not have imagined it carried to such
an extent; nor could I have imagined that the
natives would have submitted to it.

The trade of Axim is entirely in the hands of
the governor, and a fine of two ounces of gold is
levied on any native trading direct with a European vessel.   By great exertion we succeeded
in getting a little firewood on board, for which
we had to pay an extravagant price to his excel-
lency.

Departing from Axim, we arrived at Cape
Coast on the 9th, where we found the Columbine
awaiting our arrival.   She had made a good passage after parting with us off Sierra Leone, and
had been at her present anchorage nearly three
weeks.

At Cape Coast Castle we received the greatest attention and kindness from the governor,

George M'Lean, Esq. who insisted that we should take up our quarters with him while we stayed at Cape Coast. At his table we met several gentlemen connected with and resident in the colony; and certainly their appearance did not tend to confirm the generally-received opinion of the extreme unhealthiness of the climate: if they were unhealthy, they did not appear so; and if unhappy, their countenances and behaviour belied them, for more cheerful and pleasant companions I had never before met with.

On the day after our arrival, the governor treated me to a drive in his light carriage. When the reader is informed that he drove four-in-hand, he must not imagine that horses are here meant. The governor's carriage of Cape Coast Castle was drawn by four negroes, natives of the soil, who tramped along right merrily at the rate of five miles per hour. At first I was somewhat shocked at what seemed to me a little stretch of power; but discovered that the honour of being put into harness in the governor's carriage was eagerly sought for by the natives, and that those who were selected for the service were objects of envy among their countrymen. During the drive we had much conversation respecting the inter-

nal trade and our own prospects; and I shall always entertain a grateful sense of the readiness
with which Mr. M'Lean gave me his opinion and
his reasons for it,—an opinion which was afterwards confirmed by my own observation and experience.

The establishment of Cape Coast is managed
by a committee of merchants trading to that
part of the African coast : they are appointed by
government, which has a veto on the appoint-
ments.   The governor has no jurisdiction beyond
the walls of the castle ; but the natives all look to
him for protection, and his word is law over the
whole Fantee country.   It is generally allowed by
those who knew the colony while it was directly
under the control of our government, that the
fort and its dependency Accra are kept in a much
more effective state at the present allowance of
three thousand five hundred pounds per annum,
with a garrison of one hundred black men, than
when it cost the nation thirty thousand per annum, with a garrison of six hundred white men.
This is a strong proof of the credit due to the
present governor, and the applicability of the
plan to other establishments on the coast.

This colony has a direct trade with Sockatoo,

through Boussa and the Ashantee country. The Bornou kafilas bring to the eastern and northern borders of Ashantee ivory and slaves, receiving in return a few British and Portuguese goods, but principally goora nuts. Its chief export, however, is gold dust, the whole country being impregnated more or less with that precious metal, and many of the natives may continually be seen washing and sifting the gravel at the sea-side in search of it. This abundance of gold is one principal cause of the dissipated, worthless character of the people, who are, gamblers in their business, notorious liars, cowards, and thieves. Mr. M'Lean had made several efforts to induce the idle surplus population to apply themselves to agriculture, but in vain; the fascinating uncertainty of their favourite pursuit always prevailing over the sober realities of a legitimate and rational occupation. I was much interested with the account which I received of the Ashantee character: it appears to be very far superior to that of the Fantee or coast people, and it is to be regretted that we were ever drawn into the ridiculous and useless war which embroiled us with that nation.

As some proof of the musical talent of the

people, it may be mentioned that Mr. M'Lean
has a capital band formed by native musicians,
who play all the old Jacobite airs of my native
land with good taste and precision.   The leader
is the only one among them who can read a
note of music; and all the rest, amounting to
some fifteen or twenty, have learned and re-
tained them by ear.   To belong to the governor's band is the height of Fantee ambition.

The unhealthiness of Cape Coast has been
attributed to the effects of a large morass situated to windward of it; but it appears to be
much more sickly in some years than in others.
On our arrival we found the hospital closed,
and there had not been a patient in it for nine
months.   All the gentlemen resident were, and
had been for some time, in good health; in fact,
the place appeared to be getting a good character: but it must be borne in mind that there
are not twenty white men in the colony, and
none of the lower class, excepting now and then
a drunken or mutinous sailor who may happen
to be taken out of a merchant-vessel and kept
for the first man-of-war.

In the year 1824, when the fort was garrisoned

by European troops, of six hundred men, two hundred and ninety-five fell victims to the climate ; and in 1825 and 1826, of three hundred and twenty-nine Europeans admitted into the hospital, only thirty-two died.   I am inclined to think that the decrease in the mortality arose more from the survivors being seasoned to the climate, than from any decrease of the poisonous miasma.

The Fantees, as well as the Ashantees, manufacture gold into a variety of beautiful trinkets : their chains especially are highly valued, in consequence of the purity of the metal.   Mr. M'Lean showed me an Ashantee " cable-laid" chain of the most perfect manufacture, and a snuff-box which a native had made, that, with the exception of the hinges, would not have disgraced a London artist.   The usual way of getting a ring or any article made, is to weigh out the gold dust to the workman, who returns with the article completed, being less in weight by one fifth, which repays him for his workmanship and waste of metal.

The quantity of gold produced on the Gold Coast varies from forty thousand to fifty thousand ounces, and is all collected by washing.

I have seen several lumps of gold embedded in quartz, but never combined with any other mineral.

One of our seamen, who was suffering from dropsy, was landed here at his own request, and, I subseqently heard, died a few days after our departure.

CHAPTER III.

Departure from Cape Coast.—Captain Harries and an En-
gineer taken ill.—Description of a Tornado.—Anchor off
the Nun.—Death of Captain Harries and the Second En-
gineer.—Tribute to the Memory of the former.—The Ves-
sels enter the River.—Dangers of the Bar.—Preparations
for proceeding up the River.—Anchor off King Boy's Bar-
racoon.—Effects of Climate.—Remarks on the Natives.—
Continue up the River.—Anchor off Inghirami.—News of
Hostilities.—Coolness of a Native Pilot.—A Village burnt.
— The Warree Branch.—Anchor off Eggabo.—Boy and
his Wives.—Arrival at Eboe.

MR. LANDER having completed his arrangements with Pascoe, Mina, Jowdie, and some
other men who had accompanied him on his
former expedition as interpreters, we took
leave of our hospitable friend the Governor
of Cape Coast Castle at four P.M. of the 11th,
and left our anchorage under canvass. The
Columbine in heaving short broke the arm of
her anchor, and came to again with her best
bower. As we had drifted to leeward a con-

siderable distance, we continued on our course for the river Nun, expecting the brig and the Alburkah to follow us during the night.

In the course of the evening Captain Harries complained of a stiff neck, which he attributed to sleeping with one of the cabin windows open to the land-breeze. The second engineer, George Curling, was also slightly indisposed. Dr. Briggs prescribed for the latter; but Harries would take no advice, and insisted on keeping the deck.

On the following day, the 12th, we had a fine fair wind; but the Quorra under all her canvass could only make about five miles per hour, steering at the same time two and a half points to windward of our course, to compensate for the indraught into the bight. As we were drawing towards our destination, the men were now busily employed in fixing the chevaux-de-frise* round the vessel's sides. Captain Harries became worse, and symptoms of fever appeared; but we could not prevail on him to take the

* This contrivance was intended to prevent our being boarded by the natives; but fortunately we had no occasion for proving its value, although it was of considerable service to us in preventing them from crowding and incommoding our decks.

advice of Dr. Briggs. Curling, the second engineer, was rather better.

On the 13th of October the wind still continued fair, and we had remarkably fine weather. An incident happened in the morning on our meeting for prayers that produced considerable amusement at the time, ill corresponding with the gravity of the occasion on which we had assembled. Dr. Briggs, who always officiated for us, was just commencing his usual duty, when a white tornado came on us, so suddenly without its customary warning, that in an instant, being unprepared for it, the whole of us rushed simultaneously from the poop in all directions for shelter; Dr. Briggs and I made a short cut through the skylight into the cabin, while the seamen, with their characteristic love of mischief, prevented the engineers from getting below until their clothes were thoroughly drenched by the torrents of rain which accompanied the squall. On this part of the coast tornadoes are frequent; but, excepting one similar to that which we experienced this morning,—and it was not at all violent,—they always give warning;—in fact, I never saw more than three or four that did not give timely notice of their approach.

There is something awfully grand and impressive in the appearance of the heavens before a heavy tornado. A dark mass of clouds collects on the eastern horizon, accompanied by frequent loud but short noises, reminding one of the muttering and growling of some wild animal in a voice of thunder. This mass or bank of clouds gradually covers one half of the horizon, extending to it from the zenith; but generally before this a small and beautifully-formed radiant arch on the verge of the horizon appears, and gradually increases. Long before it reaches the vessel, the roaring whistle of the whirlwind is heard, producing nearly as much noise as the peals of thunder that seem to rend the very clouds apart from each other. The course of the squall is distinctly marked by the line of foam it throws up, and I have stood on the taffrail of a vessel and felt the first rush of the wind while her head sails were becalmed. The sensation it produces afterwards is cheering and delightful. From breathing a close and murky atmosphere loaded with unpleasant vapours that invariably precede the tornado, the mind becomes relieved as it were from a load, the air is fresh and clear, and everything around is exhilarating. On our voyage

we had been favoured with many of these squalls;
and when in soundings, we always anchored and
rode them out, or else turned tail and ran round
the compass with them under nothing but the
fore staysail, with the sheet secured a-midships.
With all their inconveniences, they are of the utmost benefit, as they clear the atmosphere and
render it pure and bracing, removing at once the
oppressive and noxious vapours with which it is
charged. I remember once hearing an old seaman observe respecting them, "Ay, it's only
ould Nature sneezing;" and I thought the remark
not badly applied.

Harries was much worse to-day, and still persisted in refusing either medicine or advice, from
an opinion that starving himself was quite sufficient. George Curling, the engineer, became
much better and appeared on deck.

On the 14th, poor Harries became still worse,
and at length submitted to have a blister applied on the back of his head. This was done
by Dr. Briggs, who also prescribed for him. I
did not consider him in any danger; but from
some conversation we had together I found that
he, poor fellow! thought otherwise, and for the
first time in the voyage I was affected at the

thought of losing a companion.   He talked of himself and his illness in terms of great anxiety; and I found that he had a presentiment on his mind which he could not shake off, that his visit to Cape Coast would be fatal to him.   Perhaps the remembrance of the sickness he had witnessed there in his Majesty's ship Esk, when he was master of that vessel, may have given rise to it. He desired me to take charge of the Quorra, as he did not consider his mate capable of keeping the crew in order.   Such a communication from him being so entirely unexpected, fell like an electric shock on my mind, and produced feelings of sorrow and anxiety,—the former from the idea of losing him, and the latter from my deficiency in the knowledge of seamanship, which was increased from a consciousness that the vessel was difficult to manage under canvass.

Early in the morning of the 15th, being in twelve fathoms water and a tornado coming on, I anchored, and had barely time to get the sails furled before it broke upon us and continued for two hours with great violence.   We had to veer sixty fathoms of chain before the vessel was brought up.   The weather being very thick, and suspecting we were embayed, I remained at an-

chor until twelve o'clock, when it cleared away,
and I obtained a meridian observation by which
I found that we were abreast of the river Dodo,
the current having set us into the bight. I
therefore got under weigh; but finding the swell
still setting us in-shore, we were obliged to bring
her up again in five fathoms water.

On the 16th, Harries was much worse and became delirious; Curling also had a relapse and
was worse;—though neither of them would have
been considered dangerously ill in England. At
day-break, there being a light land-breeze, we tried
the vessel again under canvass, but could not get
her ahead. At nine we got up steam, and at five
P. M. anchored off the mouth of the river Nun,
in four and a half fathoms. On our arrival I was
much disappointed at not finding the Columbine,
as our fuel was expended. In the afternoon
Captain Harries and Curling were very ill, and
both of them delirious.

At day-break of the 17th, a black pilot, named
Dedo, came on board in a canoe paddled by
twenty men, and offered to take the vessel into
the river : but having heard a bad account of
the Nun pilots, I thought it safer to take the gig
and examine the bar, at the same time allowing

him to remain on board.   I had been up all
night, attending on poor Harries; but the duty
being urgent, I departed in the gig, with four
Kroomen and a leadsman, for the mouth of the
river.   We found that the reefs extended out
two miles from the main land, and that there
were two fathoms on the bar at low water.   We
proceeded some miles up the river, and found the
brig Susan, of Liverpool, loaded with palm-oil,
and waiting for assistance to enable her to quit
the river.   She had been lying there seven months,
the last four of which she had been ready for sea,
but was unable to get out from want of hands,
having lost seven men out of twelve.   The vessel
was in a most deplorable condition, and to all appearance not seaworthy; and the miserable remnant of the crew were more like spectres than
men.   Her commander informed me that he had
purchased Mr. Lander's journal from King Boy,
and had paid two hundred pounds for it; but from
the fellow's looks I much doubted his assertion.

By two in the afternoon I returned to the
Quorra, having suffered much from the intense
heat of the sun, to which I had been exposed for
above eight hours.   My black friend the pilot,
I found, had been excessively uneasy during my

absence; but after being paid for his detention, he recovered his self-possession and left us with the promise to return in the morning. To my great astonishment, he paddled his canoe directly through the breakers. How he managed it I know not; but I certainly have seen the long canoes of these natives live in a surf that would have swamped any European boat : perhaps the number of paddles which they have, and the dexterous manner of using them, may in some measure account for it.

At four in the afternoon the Columbine hove in sight. I sent word to her commander, as she neared us, to bring his vessel up in six fathoms, with the river open to the N. E. On his arrival we found that he had missed the Alburkah the same night that he left Cape Coast.

Captain Harries became worse, and Dr. Briggs appeared very anxious about him.

At four o'clock on the morning of the 18th, I retired to rest, worn out both in mind and body, having remained by Harries the whole night. In two hours afterwards I was called by Dr. Briggs, and found Harries dying in his arms. He expired half an hour after I went to them, without a struggle. At ten o'clock in the forenoon, the

second engineer, George Curling, died with exactly the same symptoms. At two P. M. we
committed their bodies to the deep, firing the
customary salute over that of our lamented commander. During our tedious and trying passage
out, Captain Harries had sacrificed his own comfort so much to us, and endeared himself to us
by so many little kindnesses and attentions, that
his death was severely felt by me. He was much
beloved by his crew for the kindness and consideration with which he always treated them ; although at the same time he was a strict disciplin-
arian, and insisted that the duty should be done
well and quickly ; and it was gratifying to see the
sensibility which they displayed on the melancholy occasion of his death.

Considering the difficulties which Captain Harries had to contend with, he had made an excellent passage to the coast; and any seaman would
acknowledge, that the navigation of a vessel having about eight feet above the water and only
five below it was no easy matter. Certain I am
that no exertion on his part was wanting, for a
more active and energetic seaman I never met
with. I deplored his loss on my own account, as
I had a strong personal regard for him ; I la-

mented it for the sake of my copartners in equipping the expedition, for I was well aware that their interests would suffer in consequence.

At daylight of the 19th I sent the mate on board of the Columbine to get her launch hoisted out and loaded with coals for us. At noon he returned, and we immediately got up steam and towed the Columbine over the bar, on which were two fathoms and three-quarters at quarter-flood. Her long boat, which she had in tow astern, carried away her painter and was swamped in the surf. As the recovery of this boat was necessary in consequence of her great importance to us, I manned two boats and sent them after her, and after anchoring the brig in the river, went out after them in the Quorra, as the ebb-tide was now running furiously and drifting them on the lee-breakers. Fortunately we got hold of them all, but not before we had found ourselves in one and three-quarters fathom water. The tide was so strong that we could only stem it by the united power of steam and canvass, and even then were three hours in making good a distance of three miles. The sea on these bars is frightful while the ebb-tide is running. In going in we were pooped twice; but fortunately the boat was well

and steadily steered, or the consequences might
have been serious.   In the evening the Alburkah
hove in sight, and came to an anchor off the bar.

My friend the pilot did not keep his promise
to-day.   Probably the reason he did not make
his appearance arose from being kept so long a
prisoner yesterday; and as we had found our way
safely in over the bar, I thought it not improba-
ble that we should have a visit from him the
next morning.   This fellow Dedo has the reputation of being a great scoundrel, and of having wil-
fully lost several trading vessels on the bar : but
though I have no partiality for the man, it is but
justice to say that the rapidity of the ebb-tide
appears to me to have had a great deal more to
do with it than the pilot.   The ebb sets directly
across the windward bar, and any captain of a
vessel not aware of this peculiarity naturally hugs
the windward reef in going out.   If it should fall
calm, or the wind should lull for a minute, he is
lost, as the tide sweeps him instantly on the reef.
It is the nature of man, be he black or white, to
lay the blame of his misfortunes upon every one
but himself; and this unfortunate pilot Dedo has
had to bear the obloquy of all the wrecks that
have happened in the river Nun.

While entering the river I looked in vain for the battery described by former visitors as being on the eastern side of the river; not a vestige of it was to be seen, and none of the natives of whom I inquired appeared to recollect anything of it. It was rather remarkable that I had entered the river on my natal day: whether to think it a good or a bad omen I knew not, but I found myself speculating now and then on where my next might be passed. The river where we lay appeared about three quarters of a mile broad, having an average depth of six fathoms. I anchored the Columbine with both points of the entrance open, in order that she might have all the benefit of the sea-breeze.*

On the following morning I was under the painful necessity of punishing a man for striking the chief. Having made every preparation, before a lash was given I offered him the alternative of leaving the vessel for the Susan, then lying in the river; by which offer I got rid of a worthless character, without disgracing the vessel by inflicting corporal punishment.

* I was happy to find on my return, that her mortality had been much less than that of other vessels which had been in the river and had anchored within the swamps.

The Alburkah came over the bar in good style this morning, and I repaired on board her immediately to inform Mr. Lander of our melancholy loss. On reaching her I found that he had buried a man the evening before, and that the symptoms under which he had died had been the same as those of Captain Harries, and had shown themselves at the same time.

I found that King Boy was expected the following day. In the course of the day I discovered that the worthy character who commanded the Susan had been offering high wages to my men; and as I did not wish to force any man up the river against his will, I mustered the crew, and informed them that the Susan was in want of hands; that the man in command of her, as they well knew, had been offering them very high wages; and that if any of them wished to leave the Quorra, they were perfectly at liberty to do so. I found that I had not been mistaken in my calculations on the spirit of British seamen, for only two accepted the offer.*

* I need scarcely say, that when the awful mortality took place afterwards, it was most consolatory to my mind that I had adopted this measure—one which I would recommend to any person under similar circumstances.

The 21st of October was our first Sabbath in the Niger, and Mr. Jordan read the usual prayers. Mr. Lander came on board, and I was glad to hear from him that he had recovered his journal from the commander of the Susan. I wished very much to see it, as the daily journal of his adventurous voyage down the Niger from Boussa must have been very interesting; but unfortunately he had not brought it with him, and I could never afterwards get a sight of it.

At daylight on the 22nd, we commenced breaking out the main hold, preparatory to lightening the vessel of all superfluous stores before proceeding up the river. We had been frequently annoyed during the passage out by a disagreeable vapour that came from the hold, and we now found that it had been occasioned by the cocoa being stowed in bags in the provision-room under the cabin : the bags had rotted, and the cocoa had fallen into the bilge-water, and there become putrid. This I am of opinion was the principal cause of the unhealthiness of the after part of the vessel,—the two fatal cases of fever having occurred in the poop, and no severe cases in the forecastle, where the men are much more crowded together.

On the 23rd, the celebrated King Boy, the same who had ransomed Mr. Lander and his brother, and conveyed them down the river from Eboe, came from Brass Town and paid a visit to his old acquaintance. We were busily engaged in receiving goods from the Columbine, and in transhipping into her all our spare sails, &c. Mr. Lander assured me that there was no necessity for taking any salt provisions up the river, as we should get abundance before we reached Eboe. However, I thought it but prudent to take with us a supply for three months, in case of any misunderstanding with the natives, although he assigned two months as the utmost limit for our passage to and from the upper country. Mr. Lander was also of opinion that sufficient ivory would be obtained in that time to load one of the steamers; but from the conversation which I had had with Mr. M'Lean at Cape Coast, although I wished he might be right, I could not help doubting it very much.

The weather since we entered the river had been anything but favourable for our operations: it had rained almost incessantly, which was very trying to the men. My plan was to give them as much coffee as they liked to drink, and they appeared to be all in good health and spirits. Mr. Hector, our purser, also, who had been taken ill at Liberia, was now fast recovering.

By the 26th we had completed our arrangements for ascending the river, and had reduced the draft of the Quorra to five feet three inches. I was given to understand that King Boy intended to accompany us to Eboe.

At two in the afternoon we were under weigh, and advanced up the river to King Boy's barracoon, or slave-hut, situated about nine miles distant from the bar. Here I was introduced to King Forday and King Boy, both ill-looking fellows, but uncommonly civil. The latter was dressed in a Highland uniform, which had been sent out to him by my father: it consisted of a red coat, full-dress kilt, red stockings, and yellow slippers; the whole surmounted by a military hat, with a feather in it about a yard long. But he had a disagreeable, sulky cast of countenance; and with all his protestations of friendship, and expressions of readiness to serve us in any way, I was too much of a disciple of Lavater to trust him.

There were two men with them who professed
to know the channel of the river to Eboe.   By
their account there were from three to eight fathoms the whole distance.   One of them, named
Louis, appeared to be an intelligent fellow; but I
did not exactly like the look of the other.   He
asked me how much water my vessel drew: to
which I replied, two fathoms; but that, having
wheels, we could go as well upon land as
water; which piece of information he received apparenly with the most praiseworthy credulity.

The country on both sides of the river appeared to be one extensive swamp, covered with
mangrove, cabbage, and palm trees, and must no
doubt be very unhealthy.   The fen-damp rose in
the mornings cold and clammy to the feeling,
and in appearance more like the smoke of a
damp wood fire than anything I could compare
it with.   It was my practice, immediately the
men left their hammocks, to give them a cup of
strong coffee.   I had awnings stretched fore and
aft since we entered the river, to protect them
from the ill effects of the dews and exhalations.
We had also raised a canvass screen fore and aft
as high as the paddle-boxes, that, in case of being

attacked, the natives might not be able to take
their aim.

It was a subject of remark among us, and
occasioned some amusement, to see the different
effects of heat on different constitutions. Sometimes, with the thermometer at 84, I felt cold in
ఓ blanket dress; and at other times, when it was
75, I was oppressed with the heat:—it appeared,
however, to depend much on the moist or dry
state of the atmosphere. I found that a very
simple rule had hitherto kept me in excellent
health : if I felt sleepy after a meal, I considered
it a gentle hint from my stomach that I was over-
working it, and reduced my fare accordingly;—in
fact, I thought that the less one consumed the
better, as all our party appeared to have a most
unaccountable propensity to become fat. I did
not eat one half that I had been accustomed to
do in England, and yet could not keep myself
from increasing. Dr. Briggs was precisely in the
same way; and as for Lander, he was as broad
as he was long.

The natives of this part of the river appeared
to be very unhealthy. They were covered with
scabs, ulcers, and guinea-worms, and all kinds of
cutaneous eruptions, which I was inclined to as-

cribe to their mode of living. They sleep generally in the open air, they drink vast quantities of spirits of the very worst description, and their principal diet consists of various kinds of fish, from the alligator to ground-sharks. The whole country seemed deluged with water, and the miserable wretches that dwell in it, are dependent on the Eboe country for their subsistence: all their yams, bananas, plantains, and cassada are derived from thence.

We were much amused by a spirited chase which took place in the river here. A black fish, about thirty feet long, came in over the bar; and the moment he was perceived, off started a dozen canoes full of black fellows after him. They soon got hold of him with their harpoons, and after a spirited contest of two hours, in which the whole river thereabout was in an uproar with the upsetting of the canoes, the squalling and swimming of the natives, and the tugging and buffeting of these fellows and the fish, they at length succeeded in landing him. Their harpoons are about eight feet long, are loaded at the end, and are thrown with considerable force and accuracy.

In the evening I made a grand fetish, or charm, which was to ensure King Boy success against all his enemies. The process was merely dressing the vessel with blue lights and firing a few sky-rockets; in consequence of which the Quorra was always afterwards called the fetish ship, and I found great difficulty in getting any of the natives to venture on board till I had passed Eboe.

On the 27th, we were busy in our preparations for going up the river. The black pilot that had been placed on board the Alburkah by King Boy informed Mr. Lander that he had been desired by the king to take us up a branch of the river that was full of shoals, and, if possible, to run us aground. In consequence of this information, Mr. Lander went up in the Alburkah to examine the branch said by the pilot to be most eligible. It proved to have from two to three fathoms water in it, and, although narrow, was free from shoals. Both Boy and his father, Forday, were greatly enraged on our determining to attempt this channel, which, in my opinion, proved that they were meditating treachery. They both departed in high dudgeon, but promised to be on board by cock-crow the next morning.

On the morning of the 28th, however, the

cock did crow, and they had not made their appearance. We therefore commenced the ascent of the river without them, following the track of the Alburkah at the distance of a few hundred yards, in order to have sufficient time to stop or reverse the engine, in case the Alburkah met with any shoal or hidden danger, as she drew so much less water than the Quorra.

Immediately above King Boy's barracoon the river divides into several wide and shallow branches, the principal one leading to the eastward to Brass Town and Bonny. That which our pilot conducted us through appeared at first to be most unlikely to lead to the magnificent stream we were fast approaching: it ran about N.W., and for about half a mile was not more than thirty yards wide, with an average depth of about two fathoms, after which we found ourselves in comparatively more open water. The navigation for the first ten miles was difficult and dangerous, and I must confess that I was not without suspicions of the pilot's intentions, particularly when I found that he took us through a channel in which there was only a depth of eight feet. This, however, proved to be only a bar of mud; for, directly after we had passed it, we had two and a half,

and three fathoms; and in gratitude to our
pilot, we named this narrow channel Louis
Creek.

We had proceeded about thirty miles up the
river before we saw anything like land or even
mud, the mangroves only marking the channel.
Afterwards the banks began to appear and assumed a more decided character, the mangroves* being no longer seen,—thus showing
the limits of the tide. We also discovered some
barracoons erected on cleared spots on the banks
for the shelter of the canoe-men and their cargoes, as we supposed, on their passage to the
Eboe country.

At four in the afternoon we passed some mi-
serable-looking huts tenanted by as miserable-
looking men. Mr. Lander landed and made
them a few presents. We continued under
steam until seven in the evening, and I considered made good a distance of about forty-
five miles in a northerly direction. The river
had now increased in breadth, which averaged
about three hundred yards, with a depth from

* This beautiful but deadly tree does not grow in fresh
water. See Humboldt, who supposes that it gives out a peculiar and dangerous exhalation.

four to five fathoms. We passed thirteen branches or creeks running from it, one of which leads to Brass Town, and is the same that Mr. Lander and his brother passed through. We now considered ourselves to be fairly on the Niger, and felt that we had much reason to be thankful for having escaped the plot laid by King Boy for our destruction.

In the morning of the 29th we were under weigh soon after four, and found that the river continued to deepen, but the banks were still low and swampy. The course of it was very serpentine, varying from north-west to north-east and east, and sometimes to the southward of east. We passed several small villages surrounded with cocoa-nut trees and plantations of bananas and plantains : in fact, wherever there appeared solid ground it was crowded with huts. The inhabitants all mustered in front of their houses armed with muskets, and seemed, though alarmed, ready to defend their homes. At seven A.M. we passed a branch running to the south-west ; and the river immediately widened to six or seven hundred yards, and deepened to ten fathoms. The reaches were longer, and

the banks frequently covered with long grass
a few feet from the river.

At twelve we anchored off a small town called
Inghirami. King Boy in his canoe came up
immediately after we had anchored. The chief,
who for a small present had consented to supply the vessels with wood, would now do
nothing, evidently from Boy's representations.
I must say I felt very much inclined to sink
him ; his presence, I was sure, foreboded no
good.

We remained at anchor off this village till
the 30th getting wood : and here I found the
full value of our Kroomen ; they were capital
axemen, and cut in a day about one day's fuel
for the engine. I rejoiced that Captain Harries
had decided on shipping these men, as the natives which Mr. Lander had brought from Cape
Coast were useless for this sort of work.

I had here only two of my crew slightly indisposed, and was entertaining hopes that we
should get into the upper country before any
sickness visited us, the weather being fine, with
a few occasional showers. The thermometer
(Fahrenheit) varied from 80 to 83. In the af-

ternoon I sounded across the river and found an average depth of five and a half fathoms within two yards of both banks, the breadth being about five hundred yards. I concluded from this that we should find no more shallow water unless the river spread out. The water was perfectly fresh here, although it was influenced by the tide, the rise and fall being about four feet. In the course of a walk on shore with Dr. Briggs, I purchased a goat of Lilliputian dimensions.

As we were getting under weigh on the following morning, King Boy was particularly anxious that we should stop at a village which he pointed out. There was something peculiar in his anxiety to get us to stop at several places in the afternoon, and I thought it not unlikely that we should fall into trouble soon. Indeed the busy, pertinacious conduct of this fellow kept me on the alert in meeting his manœuvres, and I felt a great desire to seize on his person and keep him on board the vessel until our return. Certainly, in the event of any mischief occurring, I was determined that if he were within reach he should be the first to suffer.

The course of the river varied from N. E.

to S.E., and once to south half east, the depth being from seven to eight fathoms. We passed in the course of the day two considerable streams running into, and two running out of, the main river. The country improved considerably in its aspect also. The splendid African oak and wild cotton trees appeared on the banks.

On stepping down from the paddle-box this morning, I ran the spike of the chevaux-de-frise through my foot. It was Dr. Briggs's opinion that the sheath of the tendon was injured, which a little quiet would soon set to rights. I adopted my favourite plan of starving to keep down inflammation. We continued under steam until nine in the evening, but found it difficult to work in the dark.

We started again at six in the morning of the 1st of November, and found the river in one reach lying to the west of south. In the afternoon, one of the crew, a boy, was taken ill with an attack of ague. In the course of the day we had a visit from the chiefs of two villages that we passed. One of them was an intelligent fellow, and complained bitterly of the Brass people not allowing him to take his palm-oil either to Bonny or the Nun. He presented me with two

goats and some plantains in return for a few
things which I gave him, and had the honour of
drinking the King's health in pure Jamaica, to
his infinite delight.   I learnt from this fellow that
the Brass people take their palm-oil; but, from
his account, what they give them for it is about
one-fourth of what they charge ships for it in the
Rivers Nun and Bonny.

King Boy kept at a respectful distance astern
of us to-day; and it was only now and then, at
the end of a long reach, that we caught a glimpse
of his flag.   At eight in the evening, our fuel
being expended, I came to an anchor.

The crew had gone to rest, and I was making
myself comfortable in my berth, when, at about
half-past ten at night, a canoe came alongside,
and a note was brought to me from Mr. Lander
by the pilot Louis.   It was briefly as follows :—

"M'GREGOR LAIRD, ESQ.

"SIR,—The Eboes threaten to attack us to-
morrow.   I would thank you to have everything
in readiness to resent an attack of whatever nature it may be.   Your obedient servant,

"R. LANDER.

"P. S. They say we shall not pass this place,
although there are not ten houses in the town."

The postscript was not of a very alarming nature; but, without disturbing the crew, I called Dr. Briggs, got all the muskets and pistols loaded, and packed the cartridge-boxes. While we were thus employed, we heard the report of several shots, and on going on deck observed the whole bush on our left, which was the right bank of the river and about seventy yards from us, in a blaze of musketry, which the Alburkah was returning. Mr. Lander hailed me and said he was going to drop down the river; but on my offering to go between him and the fire, he sent Louis, the Eboe pilot, on board. We immediately got under weigh, and the pilot placed the Quorra within pistol-shot abreast of the town. I was much amused by the coolness and self-possession of this pilot. He could speak tolerable English; so I told him, if he ran us aground, the instant the vessel touched I would blow his brains out. The fellow laughed, and opening his country cloth, showed me the butt-end of two pistols; a gentle hint that two could play at that game. I gave directions for the guns to be loaded with round and canister, and in about twenty minutes we silenced the firing on shore. At our first fire the eighteen-pound carronade strated the ring-

bolt of its breeching and was unfortunately disabled for the night. At one o'clock, the firing
having ceased excepting an occasional shot from
the bush, I sent all the men to their beds and
remained on deck. I was much pleased by the
cool and steady behaviour of my men in this
affair: Captain Harries having exercised them
on the voyage out, they worked the guns with
great facility and accuracy.

At daylight of the following day (the 2nd) the
firing from the bush recommenced, and I now
discovered that our invisible enemies (for we
hardly saw one) had two swivels. The Quorra
opened her fire on the town at six o'clock with
four four-pounders and the twenty-four-pound
swivel; but finding that we made no impression
on the mud-walls of the huts, although we had
silenced the firing, Mr. Lander hailed us, and
we agreed to land and burn the town as an example to the rest. Accordingly the gig under
my command led the way; Captain Miller in the
cutter followed, and also the launch with eight
men; while the two mates, engineers, and Dr.
Briggs were left to keep up a fire of musketry
over our heads.

As soon as we had reached within half pistol-

shot of the bank, the firing from the bush and
the town recommenced ; on which our lads, giving
three cheers, pulled for the bank, where we intended to land. The bank was of mud, about six
feet high, and on my first jump on shore, I found
myself up to my knees in it; but seizing the boatswain of the Quorra, a remarkably stout fellow,
he quickly extricated me, and we gained the
top of the bank. The natives had been annoying us with musketry all this time, although
they had done no harm ; but as soon as they saw
our heads above the bank, they retreated to the
morass behind their huts. I had much difficulty
in keeping the men from following them into the
bush : however, after setting fire to the roofs of
the huts, I succeeded in getting them into their
boats again. The Alburkah's boat with Mr.
Lander had been detained, and having a strong
current to pull against, did not join in the attack,
but came up immediately afterwards. On leaving the town, I was surprised at the sudden re-
appearance of the inhabitants, who commenced
firing on the boats, uttering at the same time
loud yells. I had returned to the Quorra, when
I perceived one of the Alburkah's men standing
under the bank on a narrow edge of mud. In

the hurry of re-embarking he had been left ashore. I pulled back in the gig and took him off; but we had a narrow escape, and it was merely owing to the form of the bank and the gallant behaviour of our Kroomen that we got clear off. By my directions, the Kroomen in the launch kept up a brisk fire, under which I pulled in and recovered the man, notwithstanding the natives were firing away within a few paces of us. On mustering the crew, I was glad to find no one hurt; which was easily accounted for, as the natives fire always from the hip and never take aim from the shoulder. The boats and oars had been hit in many places.

At nine we proceeded up the river, after getting up our anchors, and came to again off a village about three miles above our last anchorage. We were received here with open arms by the chief, who was delighted at hearing that we had been destroying his neighbours.

The next morning we were informed by the natives that three men had been killed in the skirmish, and four badly wounded; the latter having been brought to the village opposite by their friends. Dr. Briggs immediately volun-

teered to go on shore and visit these men, and
endeavour, if possible, to alleviate their suffering.
On his landing, however, the natives denied that
they were there, perhaps from fear, — or there
might have been no truth in the report.

This unfortunate affair was much to be regretted, as it appeared to have arisen entirely from
the misconception of the natives. The Alburkah
had anchored off the town in the dark, and had
fired a gun as a signal to the Quorra that she
had done so. The natives imagined that it had
been fired at them, and, on the Eboe pilot landing in a canoe, they attempted to seize him. He
escaped from them, however, and regained his
canoe; and they fired on him as he paddled off to
the Alburkah. It did not appear to have been a
preconcerted plan, as no canoes had passed us on
the river, and there is no communication by land;
so that they could not have been previously informed of the time of our arrival. On these
grounds I acquitted King Boy in my own mind,
although at first I strongly suspected him of
having been concerned in the attack.

Although the natives said that we had killed
several, I did not believe it, as the suddenness of
our landing would have prevented their carrying

off their killed or wounded, and we saw no appearance of either; and I am in great hopes that
the expenditure was confined to powder alone.
In the course of the day we were visited by the
chiefs of several villages, who, having heard of
our burning the village, came to congratulate us
on having destroyed the greatest fetish or charm
depôt between the sea and Eboe; and to secure
our friendship, they all brought us presents, and
received others in return.   A soldier's jacket was
the utmost of their wishes, and afforded us repeatedly much amusement.   On receiving it, the
first thing they did, of course, was to squeeze
themselves into it, which, being all of them men
on an immense scale, they had no little difficulty
to manage.   However, having achieved it, the
air with which they walked to the side of the
vessel to show themselves to their admiring subjects was highly entertaining; whilst these expressed their delight by loud yells of appro-
bation.

At six in the evening we were under weigh,
having several of these chiefs on board, whom we
dropped abreast of their respective villages, and
continued steaming up the river until half-past
two on the morning of the 4th, it being a fine

moonlight night until that time. At six in the
morning we were again under weigh, and at nine
A. M. passed a magnificent branch about 700
yards wide, running off to the south-west. On
crossing the mouth of it we found a depth of six,
seven, and eight fathoms. I conclude this to
be the main branch of the Niger, and that the
mouth of it is that known in the maps as the
Rio Forcados or Warree, which falls into the
Bight of Benin in latitude 5° 28' N. The current is stronger in it and it is a deeper river
than the Nun branch, up which we had passed.
If this opinion be correct, a vessel ascending this
branch from the sea will save a considerable distance, and avoid the Brass country entirely.

On passing this branch the river immediately
increased its breadth to one thousand yards, the
reaches became longer, the banks higher, and
the bush that crowned them was more frequently
interspersed with plantations of bananas, plantains, and yams. It afforded me much satisfaction to find that my crew continued healthy, —
the boy who had had a slight fit of ague was quite
recovered; and whether it was the change of
scene or not, since we had entered the river the
whole crew had gradually recovered their health

and spirits, both of which had suffered in the
voyage down the coast.

At three P. M. we came to an anchor off a
small town on the west bank of the river, named Eggabo, our wood being entirely expended.
This town was the first which I had seen, the
houses of which stood at a distance from the
river. All the towns we had passed hitherto
were built on long strips of the bank, extending
by the river-side, with a morass in their rear.
The sight of this was gratifying to us, as we
thought it a promise that we were shortly about
to exchange the low swampy banks of the river
for something like terra firma.

On the following day, I accompanied Dr.
Briggs on shore, and found the town contained
about two hundred houses, each surrounded by a
bamboo fence about nine feet high. On these
bamboos were hung a vast quantity of yams with
their largest heads downwards, the position in
which the natives preserve them. We paid our
respects to the king, who received us very cordially. He was a disgusting old fellow to look
at, being covered with a cutaneous eruption, and
apparently in the last stage of dropsy, his legs
being swollen to an immense size. Mr. Lander

had presented him with a soldier's jacket, of
which he was vastly proud ; and, jealous of his
splendid acquisition, he besought us not to be-
stow such a present on his neighbour on the opposite bank of the river.

The latitude of this village is 5° 31′ N.

In the afternoon, I landed on a sandbank on
the opposite side of the river, to try the range of
some one-pound Congreve rockets, which Mr.
M'Lean had presented to me at Cape Coast.   I
found here our old friend King Boy in his canoe,
in which he had stowed away four of his wives,
two of whom were dressed in Guernsey frocks,
a third in a soldier's jacket, and the fourth,
who was the best-looking, in her own inimit-
able glossy dress which nature had bestowed on
her, freshly polished and shining with palm-oil.
This damsel was yet but thirteen years old, according to Boy's account ; who told me a long story of
the number of muskets and pieces of cloth which
he had given for her, and concluded it with a
sigh, adding, " She no be fat enough for two
moons yet," although she was, at least, fourteen
stone weight.   While this interesting history was
related, the lovely creatures, grinning and showing their beautiful teeth, were softly ejaculating

" Rum and tabac." Such a request, coming from
such lips, was perfectly irresistible, and they accompanied us alongside of the Quorra for it, but
could not be prevailed on to come on board.
However, they enjoyed the " rum and tabac" to
their hearts' content in their canoe. Boy informed me, with much earnestness, that he had
been several times robbed at the town which we
had attacked, and that the trading canoes always
passed it in the night.   He also gave me to understand, that accounts of us had reached Eboe,
and that a good " palaver" would come down
the river to meet us to-morrow.

On the following morning, in accordance with
Boy's prediction, we were agreeably surprised by
the appearance of three large Eboe canoes, each
paddled by forty men, coming down the river
with flags flying. This was the " palaver" which
Boy had announced, consisting, as we found,
of two chiefs with their attendants, whom King
Obie had despatched to meet and welcome us
into his country. They remained on board the
Alburkah for a few hours, and then left to proceed and give notice of our approach to Eboe,
about twenty miles distant. They said that
King Obie was to present us on our arrival with

two bullocks, ten goats, and six hundred yams, for which, like all the rest of the kings who have made us presents, he will, I presume, receive double their value in return.

Having completed our supply of wood, we got under weigh at six P.M. having a fine moonlight night before us; but getting into shoal water, we came to an anchor again at ten, about five miles below Eboe, while the Alburkah proceeded. Shortly afterwards Captain Hill came on board the Quorra, saying that they had found deeper water on the left bank of the river, and that the Alburkah had come to an anchor off the town. At six on the following morning we got under weigh, and in half an hour joined her at the entrance of the creek that leads up to the town.

CHAPTER IV.

Visit to King Obie.—The Palaver.—The Visit returned.—
A Female of Eboe.—Description of Obie.—The Town
of Eboe described.—Its produce.—The Delta of the Niger.
—Its Inhabitants and their Occupations.—Oil-making and
Slave-hunting compared.

PREPARATIONS were now commenced for our
visit to King Obie. The launch and other boats
were manned by Kroomen, dressed in kilts and
velvet caps, an uniform expressly intended for
gala-days; and at 10 A. M. we proceeded on shore
in state to pay our respects to the king. Mr.
Lander in a general's uniform, with a feather in
his cocked-hat that almost reached the ground,
Mr. Jordan in a colonel's uniform, and Lieutenant
Allen in his own, led the van, and attracted so
much of the natives' attention, that Dr. Briggs
and myself almost regretted that we had not
visited Monmouth-street before our departure
from England.

Preceded by old Pascoe, Jowdie, and some

other men who had accompanied Mr. Lander on
his former journey, and who were now returning
in triumph to the scene of their former exploits,
dressed in soldiers' jackets and military caps,
blowing trumpets and beating drums, accompanied by King Boy and about forty Eboe canoes
emulating them in their discordant noises, we
advanced up the narrow creek more like merry-
andrews than sensible people; and after a row of
about three quarters of a mile, in one of the
hottest days I ever experienced, we landed at the
upper end of the town amongst a great assemblage
of people of both sexes. From our landing we
had still more than half a mile to walk, surrounded by a mob of about a thousand people
armed with all manner of muskets, spears, cutlasses, bayonets and knives fastened on the ends
of poles. The heat of the weather and the
stench of the place were quite overpowering; and
the natives' pressing round us to touch the skin
of a white man required the exercise of all our
good temper and forbearance to withstand.

On arriving at the royal residence we passed
through two outer courts, each about twenty
yards square, and were ushered into a smaller
one, three sides of which had projecting roofs

supported by rudely-carved pillars. The other formed the women's apartment, and through the doorway half-a-dozen faces were visible. These we understood were Obie's favourite wives. On one side of the court, under the roof, stood Obie's throne, covered by a grass mat of most elegant manufacture; and under the opposite verandah was an upright loom, with an unfinished web of grass cloth.

After waiting about ten minutes, a side door opened, and in rushed Obie, a tall man with a pleasing countenance, dressed in scarlet cloth. He wore a cap made of pipe coral on his head, much the same shape as the fool's cap of our schools; and thirty or forty chains of very large pieces of coral were passed round his neck and loins. He had on also a great number of armlets and leglets of the same article; indeed, I should say he had nearly one hundred pounds' value of coral on him. Poor Jordan was the first he saw, and rushing upon him, at once he gave him a most fraternal hug; then shaking hands with Lander and myself, he took his seat on the throne, placing us on each side of him.

The interview lasted about a quarter of an

hour, and I was much struck with the gentlemanly and agreeable manner of Obie. The term
'gentlemanly' may appear misapplied to an untu-
tored African negro ; but King Obie displayed
towards us the very essence of gentility, in the
most lively attention to our wants and comforts.*
After the conference he accompanied us to the
boats, walking familiarly with his arm round Mr.
Lander's neck. I could not help being amused
with Boy's conduct during our interview. This
man never spoke to Obie without going down on
his knees, and touching the ground with his head.
Boy walked down to the boat with me; and on
my joking him on his abject behaviour, he,
courtier-like, replied, "King Obie too much palm-
oil, King Boy too little." We embarked in our
boats, and on quitting the landing-place there
could not have been less than between two and
three thousand people present: we gave them
three cheers, and they, with their monarch at
their head, heartily returned them. Although
much gratified by the kind and cordial reception
we had met with, I was glad to get on board my

* I found afterwards that he was equally attentive to us
in distress.

own vessel, being completely worn out with the heat, and disgusted with the stench of the town. We received from King Obie a fine bullock, five goats, and three hundred yams; and two of his sons, both of them fine young men about sixteen years of age, accompanied us on board.

On the following morning, the 8th, we all of us felt the effects of our roasting in the sun; and the little rest which we got during the night, from the torments of sand-flies and mosquitoes, who no doubt considered our arrival as a treat, did not tend to promote our recovery. At day-break I was much pleased to see a fleet of canoes of all sizes leaving the town for the purpose of collecting palm-oil, yams, and other provisions, for it gave an assurance of the regular and industrious habits of the people. There could not have been less than from one hundred to one hundred and fifty; and in the evening they came dropping in with their cargoes of yams, bananas, and palm-oil in large gourds. It was the most gratifying proof of regular and honest industry that I had yet seen in Africa.

At one P. M. we fired a royal salute in honour of King Obie, who went on board the Alburkah

Engraved by W Finden. A.R.A. from a sketch by Mr. Mc Gregor Laird.

KING OBIE VISITING THE STEAM VESSELS.

to visit Mr. Lander. He was escorted by seven
war-canoes, each containing between sixty and
seventy men, and about fifty others containing
from one to fifty persons in each. He dined with
Mr. Lander, having brought six of his women to
fan the company during the repast; and remained
on board until four in the afternoon, when he
landed under a salute from the Alburkah. I had
allowed one half of the Quorra's crew liberty to
go on shore with the second mate. They met
him as he landed, and he instantly set to and embraced them all in their turn. He was afterwards
supported to his palace by two of the sailors,
where he regaled them with palm-wine and
roasted yams. The first mate and the engineer
of the Quorra had been on shore the day before,
and were treated in the same way: in fact, all
the natives were vying with each other in paying
them attention.

Among my visitors on this day was the largest
woman I ever beheld—and a jolly good-tempered
dame she was. On thanking her and making
her a small present for the attention which she
had shown my men while on shore, she assured me that " her belly had long been hungry to

see white men and their houses," and that she
had come alongside to see the interior of mine.
I would very willingly have got the old lady on
board, had the foreyard been strong enough;—
as it was, she was obliged to content herself with
a peep through the cabin windows. She certainly could not have weighed less than twenty-
five stone, and she informed me that she was
mistress of more than two hundred slaves, whom
she employed in collecting palm-oil, cultivating
yams, &c. Her personal charms were set off by
a straw-hat nearly five feet in diameter, about
a dozen brass bracelets, two ivory leglets ten
inches wide by about six deep, and necklaces
innumerable. I promised her on leaving us
to visit her on the following day, and was fully
determined to do so.

Accordingly, on the 9th, at seven in the
morning, I landed with Dr. Briggs, to pay a visit
to my fat friend, and to have a look at the town.
We found the lady at home, surrounded by her
women, and had a long conversation with her, as
we found her an intelligent, clever woman. We
learned that she owned several canoes, which she
employed trading on the river, both above and
below Eboe. After staying some time with her,

we paid Obie a morning visit, and were kindly
received by him. He showed us his favourite
wife, a fine woman of about sixteen, and jet-black,
which I agree with him is preferable to any of
the intermediate shades. Her apartment was
remarkably clean, and the walls of it were lined
with matting. Obie himself is a particularly
handsome man; he stands above six feet high,
has a prominent nose and oval face, good eyes,
and a pleasing expression of countenance, com-
bining intelligence with good-nature. I asked
him the distance from Brass Town and Benin:
his answer was, Brass Town, seven days; Benin,
four days. From his fondness for coral ornaments, I should imagine him to be of Benin
extraction : he is certainly of a very superior
breed to any we have yet met with. After a long
conversation, in which he naturally expressed his
surprise that we were not going to trade with
him for palm-oil, he presented us with a fowl
and some plantains, and we took our leave highly
pleased with our visit. Obie is decidedly the
most intelligent black man I have yet met with,
and far superior to those on the coast. On our
return through the town, we met a representation
of the devil, (white of course,) carrying before

him a human skull covered with blood, and followed by a mob of about a hundred fellows armed with all manner of instruments, from a musket to a penknife. His Satanic Majesty was particularly civil to us: perhaps, being of the same colour, he looked on us as his brethren.

The town of Eboe stands by the side of a creek running parallel with the Niger, and in the flooded season communicating with it at both ends. On a rough estimate, the town consists of eight hundred to a thousand houses; and allowing on an average six people to a house, will give the amount of population, two-thirds at least of which may be considered as under fourteen years of age. The inhabitants are the most enterprising and industrious traders on the Niger.* The town itself with its immediate vicinity is unhealthy, owing to the swampy nature of the ground: we found but few old people of either sex, and a great number of young men who appeared debilitated and aged.

The staple trade of Eboe consists of slaves and palm-oil. The value of the former varies according to the demand on the coast; but the

* As a proof of which, I may state that I have met with them as high as Fundah.

average value of a lad of sixteen may be taken at sixty shillings, and that of a woman at something more.

Palm-oil is produced in immense quantities about Eboe, and is collected in small gourds, each capable of containing from two to four gallons, from which it is emptied into trade puncheons. Some of these, belonging to vessels in the Bonny, I saw in canoes at Eboe; but, generally, the gourds are taken in large canoes to a market-place on the Bonny branch of the Niger; which branch being dry in the dry season, the Eboe oil then finds its way through the Brass creek to the Bonny. As I did not purchase any oil at Eboe, I cannot with accuracy give the price; but I consider it may be procured in any quantities there at four pounds per ton. Little ivory passes through Eboe, the greater part going overland to Lagos and Whydah, or down the Bonny creek or Benin branch, both of which flow out of the Niger above the town. If a regular trade were established with Obie, (and no man can be more anxious for it,) I have little doubt that he would soon stop up those channels, and monopolise and concentrate the trade. No man has greater power to do this than he has, situated as

he is at the head of the three great outlets of the
Niger,—the Benin, the Bonny, and the Nun.

Between the head of these outlets and the
sea, a distance of one hundred and sixty miles, is
the delta of the Niger, which delta extends from
Lagos on the west to Old Calebar on the east.
Through this delta its waters are discharged into
the ocean by twenty-two mouths, the principal
of which are the Benin, Warree, Nun, Bonny,
and Old Calebar.* The Nun branch is the only
one yet explored, and a general description of it
may not be misplaced here.

The course of the Nun branch between the
sea and Eboe is exceedingly serpentine; but its
mean course from Eboe to the sea is south half-
west by compass. Its breadth varies from
thirty yards at the entrance, or Louis Creek,
as we named it, to one thousand or twelve
hundred at Eboe, gradually widening as it nears
its parent stream. An astonishing number of
branches of all sizes flow out of it, the larger
ones all in a south-west direction; and on exa-

---

* Perhaps Old Calebar should not be called a mouth of
the Niger, although communicating with it, as it most probably receives a great portion of its water from the eastward.

mining the chart, five large rivers, more or less
navigable, will be found between the Benin and
the Nun rivers. The branches to the eastward
are all of them small, being only navigable by
canoes, and intersect the country in all directions, forming a labyrinth whose windings and
extent we are fortunately not required to ex-
plore. The whole of this country within the
influence of salt water is covered by mangroves,
and above the influence of the tide by a dense
mass of vegetation, principally composed of varieties of the palm species. Its population is
scanty, and confined to the banks of the principal streams, which from the force of their current in some places throw up a bank of sufficient
consistency to support the mud-huts of the natives ; and wherever this is the case, the bank
is crowned by a village containing from one
hundred to one thousand inhabitants. The total
population between Eboe and the sea on the line
of the Nun branch does not exceed four thousand adults, if so much. The women and children are employed in collecting palm-oil; the
men, in trading to Brass and Eboe, kidnapping
their neighbours, and drinking the worst description of spirits.

The sovereignty of the river belongs to the most powerful for the time being. The chief of a village of one hundred inhabitants may and does war with that of his neighbours on the opposite bank of three times the amount of population. But theirs are bloodless wars—a human being is too valuable to be shot; he is caught and expires in the hold of some slave-ship, or, after the purgatory of a voyage, becomes the blessed inhabitant of that negro paradise which, if we believe a sugar-planter, is only to be found in a cane plantation.

The collection of palm-oil is lazily and indolently followed. The trouble of catching a man is trifling—that of manufacturing a ton of oil trifling still, but considerably more ; the price of both is about the same : can it be wondered at, then, that the production of oil does not increase more rapidly ? It is some matter of surprise to me that where there is such a brisk demand for slaves it is manufactured at all. The capture of a man partakes of the exhilarating nature of a hunt, while the collection of palm-oil is devoid of excitement, and becomes the sober tedium of business : this affords nothing to satiate the appetite for cruelty common to man in his natural uncivilised state—that affords him ample gratification, for slave-hunting is pre-eminently cruel. Of all the baits which have been employed by the great tempter of mankind, that of buying and selling flesh and blood has been the most subtle and successful; and if it has triumphed over the minds of enlightened and conscientious men in our own country, can we wonder that it is irresistible among savages? I am so assured of this, that I feel convinced if an African were offered double the reward for the same amount of labour that he employs in kidnapping a man, for any honest employment, he would prefer that of catching the slave.

After stating the pursuits of the inhabitants of this dreary region, it is almost unnecessary to say anything of their moral qualities: a people so thoroughly debased, demoralised, and degraded, I could not have conceived existed within a few miles of ports which British ships had frequented for a century. But it only adds another to former proofs, that the intercourse between civilised and savage nations has hitherto been productive of anything but good to the latter: it has robbed

them of the few good traits their character possessed, and engrafted on their naturally vicious
dispositions the worst qualities of civilised life.

If their moral constitution be bad, their physical one might have been given them to match it,
living as they do on the lowest vegetable and
fish diet, when they are fortunate enough to procure it.   They are frequently obliged to subsist
on vermin and garbage of all kinds; and, as a natural consequence, their bodies present the most
disgusting appearances, from ulcers, and the accu-
mulated filth of personal uncleanness.   This is
the general character of the people inhabiting
the marshy ground, as we found them; nor was
any improvement observable in them until we
came to Eboe.

The quantity of palm-oil at present made on
the banks of the Nun is not, in my opinion, one
twentieth part of the natural production.   I have
seen the palm-oil nuts of the season lying on the
ground neglected as an article too plentiful to be
valuable, and am convinced, that in case of any
efficient stop being put to the slave-trade,
the production, if the demand continued, would
be four times as much in as many years.
The very creeks and branches that perplex

and endanger the European would be frequented by the natives, in a peaceable state of the
country, as so many channels of communication between the palm-forests and the principal
ports. There is no doubt that they were intended by our benevolent Creator for the purpose
of facilitating the communication in a country
where, from the nature of the soil, land-carriage
is impracticable.

However the disorganised state of the country
may injure its inhabitants, it was perhaps of advantage to us, as it prevented any combination
among the villages to impede our ascent of the
river, and explains the feeling which prompted
the chief of the village immediately above that
which we attacked and burnt, to congratulate us
on our arrival at our success. A system that
creates universal distrust is not the most favourable to the development of combined plans of
operations against a common enemy. But I have
before stated that the skirmish originated in a
mistake, and have merely alluded to it here to
remove the impression generally entertained of
the natives on the Nun branch of the Niger. *

* The much-lamented death of Mr. Lander and some of
his companions is no proof against this assertion. A man

with valuable property in his possession, without the means
of defence, may be, and constantly is, exposed to robbery
and assassination in the most civilised countries of the world.
Mr. Lander was ascending the river in an open boat, with a
quantity of cowries, the specie of the country; his boat
grounded on a bank near a small town; the temptation of
plunder was too great for the natives to withstand, and he
was sacrificed.   But if Mr. Lander had been on board his
steam-vessel, the result would have been very different: it is
very unlikely he would have been attacked at all.

CHAPTER V.

*Departure from Eboe.—The Quorra aground.—Mr. Jordan
taken ill.—The Quorra's Crew attacked with Fever.—Proceed
to Damuggoo.—Awful Mortality on board the Quorra.—
Arrive at Attah.—Reflections on the Mortality.—Situation
of Attah.—Its advantages.—Natives' method of killing the
Alligator.—Depart from Attah.—Messenger from the
King.—Difficulties of the Navigation.—Proceed to Bocqua.
—The Kong Mountains.—River Scenery.—The Quorra
aground.*

AT eight o'clock in the evening of the 9th of
November, we left Eboe, and by the light
of a splendid moon threaded our way through a
very intricate navigation until two o'clock in the
morning of the 10th; when, being in very shoal
water, we came to anchor about fifteen miles from
the town. This was the widest part of the river
that we had yet seen,—its breadth cannot be less
than three thousand yards; and here the river
throws off its great branches,—the Benin and

Bonny.* We crossed the Benin branch, and found it about eight hundred yards wide, with a depth varying from ten feet to six fathoms. The Bonny branch appeared to run out in a south south-east direction; but as we were some distance from it, I cannot speak from personal knowledge of its size and depth. On the sandy point that formed its entrance were a great number of temporary barracoons, and nine or ten canoes, one or two of which had trade puncheons in them: they were evidently trading for palm-oil, the bank being covered with immense quantities of the gourds in which the natives collect it.

The river after throwing off these branches appeared about fifteen hundred yards wide, divided by sandy islands which were overgrown with long grass. The marks on the banks would indicate these to be entirely covered when the river is swollen. About eight in the morning, we passed a branch flowing in from the north-east, but apparently shallow. It is about four hundred yards wide, and above it are several populous villages.

At two in the afternoon, in crossing the river, we ran aground. Fortunately we were only going

* What I presume to be those rivers.

at half speed.   The engine was immediately re-
versed, and in half an hour we backed off, but
not before getting the carriage-guns and chain-
cables aft on the poop, to bring her down by the
stern.   While aground, we had three and a half fathoms under the stern, and only three feet under
the bows; the bar of mud and sand being worn
almost perpendicular on one side by the current.
In the course of the afternoon the vessel was
surrounded by canoes, with goats, yams, plantains, and bananas for sale.   On firing the evening gun, as usual, it was amusing to see the
general rush for the shore : they seemed perfectly
to understand the foxhunter's motto, " Every
one for himself, and God for us all."

At ten A. M. of the 11th, we came to an
anchor, close to the western bank of the river,
in eight fathoms.   I was sorry to hear that Mr.
Jordan, in the Alburkah, was attacked with
symptoms of fever.   He was immediately removed
by Dr. Briggs into the Quorra; and in the course
of the evening the Doctor reported to me that
four of my own crew were attacked with the same
symptoms.   They were apparently of a slight
nature; but, after witnessing poor Harries's illness, I became alarmed at anything.

On the 12th, I sent the Kroomen on shore to cut firewood. The bank abreast of our anchorage was about ten feet high, and formed of a stiff clay: the country about us was not very marshy, and but thinly wooded. At eleven A. M. of this day I was seized with fever. The symptoms in my own case were a severe throbbing headache, a burning pain in the feet and hands, and a deadly sickness at the stomach; all of which symptoms attacked me simultaneously, and without any previous indisposition. Dr. Briggs in the course of the day was also attacked in the same manner, besides nine more of the crew; and Mr. Jordan continued much in the same state.

In the afternoon, Mr. Lander came on board, with the brother of the King of Damuggoo, who was on his way to Eboe market with sixteen elephant's teeth and a number of slaves. So sudden was the prostration of strength in my case, that, though perfectly well a few hours before and quite sensible, I could hardly crawl on deck to see him. The man left the ivory on board, and proceeded on to Eboe with the slaves. We were to pay for the former on our arrival at Damuggoo, and the confidence which was shown in this act was both unexpected and gratifying.

At eight in the evening, about nine hours after my attack, I was much relieved by copious vomiting. Captain Hill, by an observation, made the latitude of this deadly spot 5° 54′ N.

On the following morning, the 13th, I found myself much better; the fever had left me as weak as a child; but I found only one white man and a mulatto of all the crew able to do their duty. By the evening, the Kroomen had got all our wood on board; and I was in hopes that a change of scene on the following day would produce a good effect on the crew. They were not absolutely ill, but complained of languor—had a disinclination to move, a feeling of nausea, and excessive prostration of strength. I tried in vain to cheer them: it seemed as if a deadly gloom hung over the vessel. Dr. Briggs was in a high fever, but sensible, and persisted in attending the men. I prayed, I trust sincerely, to my Maker, that, however my body might be affected, my mind might be spared.

From this date to the 5th of December my own journal is a blank, which from that of Captain Hill and my own memory I have only been enabled to fill up. I may premise, however, that Captain Hill's journal extends to our arrival at

Damuggoo, that my delirium only lasted a few
hours, and that I have a most vivid recollection
of the dreadful scene which our unfortunate vessel presented.

On the morning of the 14th, we left our anchorage, in company with the Alburkah; and
from my hammock on deck under the awning I
gave the necessary directions to Harvey, the
only European able to do duty. About eleven
o'clock, this poor fellow was taken ill with the
same symptoms as the rest; and from the excitement, or some other cause, I became much
worse. I managed to hail the Alburkah, and
stated that I had no one on board who could take
charge of the vessel, desiring at the same time
that Captain Hill might be sent on board to take
charge. Although much against his inclination,
he came and did so.

At four P. M. our wood being expended, we
came to an anchor; but as we found the wood
there of a bad quality, we took on board only a
small quantity.

At two P. M. on the 15th November, we again
came to an anchor, in a more convenient situation, and where the firewood proved to be of a
better quality. Captain Hill was here seized

with fever, and obliged to return on board the
Alburkah. His was another instance of the
power of the imagination: he had always had
a great dread of the Quorra,—and certainly she
was by far more unhealthy than his vessel, the
Alburkah.

On the 16th, Mr. Lander, who still enjoyed
good health, came on board, bringing with him
a mulatto who had acted as fireman on board the
Alburkah to work our engine, the whole of my
crew being confined to their hammocks. We
then got under weigh, taking the Alburkah in
tow, and at length arrived at Damuggoo.

On the 18th, Mr. Andrew Clark, a fine young
gentleman about eighteen years of age, died.
He had joined the expedition as a volunteer
against my wishes, but with the full approba-
tion of his friends, with whom I was intimately
acquainted. Poor fellow! he expired with the
utmost calmness, drinking a cup of coffee; and
his amiable and obliging disposition having endeared him to the crew, his death threw an
additional gloom of despondency over these ill-
fated men. In the afternoon James Dunbar,
one of the firemen, died.

On the 19th, my chief mate, Mr Goldie, and

my sailmaker, John Brien, followed; and on the morning of the 20th, our supercargo, Mr. Jordan, expired. I thought at the time that Dr. Briggs had died also; as, while he was endeavouring to revive Mr. Jordan, he swooned and remained insensible for a long time. In the evening of the 20th, Mr. Swinton also died:—he was a most respectable man, and filled the situation of carpenter; he was a native of Grangemouth, and having been a resident many years in the Indian Archipelago, thought that no climate could affect him. A few hours after his death, Mr. Millar, our chief engineer, a young man of high promise and respectable connexions in the South of Scotland, followed him.

On the 21st November, we lost William Ramm, the steward; William Parry, an apprentice; and Gardner, a seaman. On the 22nd, these were followed by William Ellison, the second mate, and a fine lad about sixteen years old whom Captain Harries had picked up and adopted in Dublin: his name was George ———, and I believe he was respectably connected, and entitled to some property when of age. The 23rd of November was a day of respite; but on the 24th, Hugh Cosnahan, a seaman, died, and for another interval the mortality ceased; the Quorra having lost thirteen, and the Alburkah two men. Two or three days having passed without a death taking place, allowed us to hope that some of us might survive the voyage; but at first, from the mortality occurring so rapidly, few of us expected to be spared to tell the melancholy tale. The latitude of Damuggoo, where our companions lie buried, is 6° 31' N.

On the 27th we arrived at Attah, Mr. Lander in the Quorra still towing the Alburkah. On the 28th Mr. Lander visited the king, who received him, I understood, with more than usual pomp.

On the 5th of December I was able to write in my journal, and I find the following entry :—
" Recovering slowly, but I hope surely. I have living, Alexander Hector, purser; John, the second steward; Harvey, Kirby, Belfrage, and Davies, seamen; all as weak as myself, and crawling about the decks more like spectres than men.

" ' A mother had not known her son,
Amid the skeletons of that gaunt crew.'

" Dr. Briggs has been closer run than myself, and for two nights I expected every moment would be his last : he is now, thank God! doing well. Last night, about ten o'clock, Johnson, a mulatto fireman, died; and in the morning Captain Robert Miller was found dead in his bed: he had expired so calmly, that no one was aware that his spirit had fled. Johnson died of epilepsy succeeding his fever, which had been but slight. Mr. Lander buried them both on the west bank of the river."

As the awful mortality detailed in the preceding account has few parallels, and though aware that the opinion of a non-professional man is of little importance, yet, being placed as I was in a situation of no common responsibility, I consider it right to state, first, the precautions taken to prevent sickness; secondly, the predisposing causes from which I think it originated; and, thirdly, the course which the experience I have acquired enables me to recommend to persons in similar expeditions.

First.—The precautions taken were, (with the advice and concurrence of Dr. Briggs,) immediately on entering the river to spread the double awnings fore and aft the vessel; the sides of the vessel were raised by canvass to a height of eight feet from the main deck; chloride of lime was daily sprinkled in the forecastle and cabin, and

upon the decks. White men were never allowed
to work from under the awnings, and on no account permitted to sleep exposed to the dew.
Their diet consisted of the navy allowance, coffee
being substituted for cocoa every alternate day,
and fresh meat when it was to be had; their
allowance of spirits was one glass of rum per day;
those who preferred it had brandy in its place;
no palm-wine or other intoxicating liquor of the
natives was allowed to be brought on board, and
little if any was smuggled; in addition, they had
a cup of strong coffee served out to them at daybreak. A few of them were on shore for a few
hours at Eboe, in charge of the mate; and, with
that exception, they were altogether on board
their vessel.

Secondly.—The principal predisposing causes
were, in my opinion, the sudden change from the
open sea to a narrow and winding river, the
want of the sea-breeze, and the prevalence of the
deadly miasma to which we were nightly exposed
from the surrounding swamps. The horrid sickening stench of this miasma must be experienced
to be conceived: no description of it can con-
vey to the mind the wretched sensation that is
felt for some time before and after daybreak.

In those accursed swamps, one is oppressed, not only bodily, but mentally, with an indescribable feeling of heaviness, languor, nausea, and disgust, which requires a considerable effort to shake off. Another cause of sickness was the want of excitement. We had been led to expect great opposition on our passage to Eboe, and the minds of the crew, in expectation of it, had been raised to dare anything, and to overcome every obstacle to our progress: with the exception of one slight skirmish, we met with none, and instead of the excitement produced by a sense of personal danger, which would in a great measure have prevented the men from thinking of the unhealthiness of the climate, they had before them, every day, a tedious, uninteresting, and protracted voyage through a flat and marshy country. It may be strange that the absence of danger should be assigned as a predisposing cause of sickness; but I feel confident that experience will bear me out in the assertion.

Thirdly.—To any persons under similar circumstances, I would recommend to keep up the excitement of the crew in every possible way — by giving them, when in the swampy country, triple the usual allowance of spirits; to feed them as much as possible on fresh provision; to employ them constantly, never allowing them to be idle for a moment; and, above all things, to have music on board. My reason for giving this advice is, that I consider a European in that climate as placed in an unnatural position, and that the only chance there is of saving him is by stimulating him to the utmost, and preventing his mind from sinking into that state of despondency to which both the country and the climate so fatally predispose it. As change of scene is also a powerful stimulant, the greatest exertion should be made to reach Eboe, which, with coals for fuel and a fast steam-vessel, may be done in forty-eight hours. I am aware that this stimulating system cannot last; but if it lasts long enough to take the crew alive through the swamps into the upper country, it will have effected all that is wanted, and can be gradually relinquished.

Mr. Lander's behaviour during our sickness did him infinite credit. He was indefatigable in his attention to the sick, and bled and blistered the men as if he had been a regular licentiate. The application of a blister over the head seemed to give the greatest relief, and in my own case and that of Dr. Briggs was, under Providence, the means of saving us. After this is done, I do not think that medical skill can go further — the question becomes a pitched battle between the fever and the constitution of the person attacked with it; and medicine, after the first necessary emetics and purgatives, does more harm than good.

In recovering, the greatest caution is necessary; excitement of any kind must be avoided, and great care must be taken that the patient does not give way to his voracious appetite: a relapse is generally fatal, and always more dangerous than the first attack.

The town of Attah, off which we had anchored, presented a most picturesque appearance. It is seated on the summit of a hill, the perpendicular side of which rises immediately from the river to the height of about two hundred and fifty or three hundred feet. The King of Attah has the reputation of being the most powerful between the sea and Fundah, and carries on a considerable trade in slaves and ivory. We found two men there from King Peppel of Bonny, buying slaves for their master: the price was as low as five dollars, or goods of that value, for a prime slave.

Mr. Lander, accompanied by Captain Hill, went on shore several times, and described the view from the town as most splendid. The Kong mountains are seen in the distance stretching from west north-west to south-east, and from their tabular appearance, I apprehended, would prove to be of the trap formation. The hill on which Attah stands appears to be some conglomerate,* and forms the extremity of a low range of hills which constitutes the eastern boundary of the river. The appearance of the western bank is strikingly beautiful, and I anticipated much benefit to all the survivors by the change from the low and monotonous scenery. An observation by Captain Hill placed Attah in lat. 7° $6\frac{1}{2}'$ N. It is healthy, and the only place we have yet seen in the river where a European could possibly exist for any length of time. It has many natural advantages, and on some future day will be a place of great importance. Situated as it is above the alluvial soil and at the entrance to the valley of the Niger, it commands at present the whole trade of the interior ; which trade, although trifling at

---

* It is very similar to the volcanic bluffs at Clarence Cove, Fernando Po.

present, it requires no prophet to foresee will at some time hereafter be immense. The inhabitants of Attah are enterprising traders, and monopolise in a great measure the trade above the town. Notwithstanding this, we had been lying at our anchorage for ten days, and could see no prospect of opening any trade with them; not from indisposition towards us on the part of the natives, but from their dilatory habits — time being of no importance to them. Perhaps another reason might be, that, with the exception of slaves, they have little to trade with.

One day, while we lay at anchor off the town, I witnessed one of the most ingenious ways of killing an alligator that could be imagined. One of these huge creatures was discovered basking on a bank in the river, a short distance ahead of our vessels. He was observed by two natives in a canoe, who immediately paddled to the opposite side of the bank, and having landed, crept cautiously towards him. As soon as they were near the animal, one of the natives stood up from his crouching position, holding a spear about six feet long, which with one blow he struck through the animal's tail into the sand.

A most strenuous contest immediately ensued; the man with the spear holding it in the sand as firmly as his strength allowed him, and clinging to it as it became necessary to shift his position with the agility of a monkey; while his companion occasionally ran in as opportunity offered, and with much dexterity gave the animal a thrust with his long knife, retreating at the same moment from within reach of its capacious jaws as it whirled round upon the extraordinary pivot which his companion had so successfully placed in its tail. The battle lasted about half an hour, terminating in the slaughter of the alligator, and the triumph of his conquerors, who were not long in cutting him into pieces and loading their canoes with his flesh, which they immediately carried to the shore and retailed to their countrymen. It is evident that the success of this plan depended on the nerve and dexterity of the man who pinned the animal's tail to the ground; and his contortions and struggles to keep his position were highly ridiculous and entertaining.

On the 5th of December, Mr. Lander, accompanied by Captain Hill, went on shore to inform the king of our intention to leave him on the following morning; but after a fruitless attempt to see him, they returned on board, and decided on proceeding up the river without waiting to take leave of him, at daylight. During the day some ceremonies were performed ahead of the vessels in the river to prevent our passing upwards, his Majesty of Attah being very desirous that we should not proceed further up the river. These ceremonies were performed by men in canoes, fantastically dressed. They stood up and made a variety of expressive gestures, and concluded their proceedings by throwing into the river what appeared to me to be alligator's flesh.

On the 6th of December I found myself much better; but Dr. Briggs had a smart return of fever, brought on by an attempt which he made to attend to some of the convalescents. Captain Hill and Mr. Lander came on board at six A. M. and got the Quorra under weigh, with the Alburkah in tow. The engine was worked by a mulatto belonging to the Alburkah, the only remaining fireman of both vessels. As the vessels gathered way and passed the place where the fetish or ceremony of yesterday was made, the natives, who were crowded on the hill on which the town stands, testified their astonishment by a general yell.

The scenery of the banks, but particularly the
western one, about this place is beautiful.   We
passed one of the loveliest little towns I have yet
seen, which I called the Crow's Nest, from its
being seated on the summit of an almost perpendicular rock about one hundred and eighty feet
high, and apparently of basalt ; but it was so overgrown with vegetation, that the face of the stone
was scarcely visible.   We found the navigation
difficult, the current stronger than below Attah ;
which we might have anticipated, as we were
now entering the valley of the Niger.   The level
of the river appeared to be rising.

After a few hours' steaming, we anchored in
a favourable position to obtain wood.   Captain
Hill had a return of fever this afternoon.   It is
something remarkable that this vessel, the Quorra,
seemed to make every one ill except Mr. Lander,
who continued in excellent health.   Captain Hill
returned on board his iron vessel, the Alburkah,
vowing that he would never leave her again.

In the evening a canoe arrived from Attah,
with his majesty's brother, stating his surprise at
our leaving so abruptly, and repeating his protestations of friendship, with the assurance of his
willingness to trade if we would turn back.   I was

much amused with the fellow's cunning, and the ingenuity he displayed in varying his lies; at one time pretending that he was unwell, at another that he was consulting his gods—always ending with "Why won't you take men?" (instead of ivory.) However, neither he nor his errand had any effect with us.

On the 7th we were under weigh at noon, in hopes of reaching Bocqua before night; instead of which we were obliged to come to again about three miles from our last resting-place, having spent the whole afternoon in fruitless endeavours to find the right channel, the river being divided by sandy islets into a number of small channels. By the marks on the bank under which we anchored, the river appeared to have fallen twenty feet.

On the following morning I found myself sufficiently recovered to take charge again of my own vessel; and my remnant of crew were gradually and, I hoped, surely recovering. After leaving our anchorage, we were obliged to anchor again three times, while the channel was examined by a boat; and after making good about twelve miles, we came to an anchor opposite a beautiful little town, situated as usual on the top of a small hill. The least depth of water we had was six feet, and the next cast of the lead was four fathoms. The bed of the river appeared to be formed by ridges, as we frequently found ten feet, and the next moment six and seven fathoms. This reach of the river runs north-east by compass; but it was nervous work for a convalescent to navigate a vessel under such circumstances.

The scenery which we passed to-day was of the most enchanting description. A succession of flat table-hills with perpendicular sides were backed by the Kong Mountains in the distance, appearing to rise to the height of between two and three thousand feet. The summits of all were of a tabular form.* To us who had been for the last two months in a flat, marshy, and monotonous country, no doubt they appeared more interesting than they really are; but I shall never forget the feelings with which I surveyed their hazy outline, though so weak at the time as to require two men to lift me up to behold them. Dr. Briggs continued very unwell, but I gained strength daily. On hailing the Alburkah, I found that Captain Hill and

---

* I afterwards found two or three remarkable hills of a sugar-loaf shape.

Lieutenant Allen were both in a very precarious state. The name of the town off which we had anchored is Attacolico.

On the 9th, Dr. Briggs, to my great relief, became better. We read prayers in the morning, for the first time since the mortality, to the miserable remnant of our crew. We formed a small, but, I trust, a sincere congregation.

In the morning of the 10th, Mr. Lander's servant went on shore and shot three guinea-fowls, which proved excellent eating. He informed me, on his return, that the natives were much annoyed by wild beasts, which committed such depredations that they found it impossible to rear either goats or fowls. The village appears to be newly formed, and Pascoe informs me that the settlers have been driven from the opposite bank of the river by the Kacunda people. In the afternoon, Mr. Lander took the gig, intending to examine the channel as far as Bocqua, and to return on board in the morning.

On the 11th, in the morning, a great number of canoes passed us returning with slaves and ivory from Bocqua market to Attah. From some of them I purchased eleven scrivelloes, but paid more for them than would have been asked on the coast. I am told that the fair or market is held every ten days, and is attended by Eboe and Attah people from the low country, and by Rabbah and Fundah traders from the north.

In the afternoon Mr. Lander returned, having been unable to reach Bocqua until this morning. He had slept a few hours on a sandbank, daylight being necessary to enable him to find the proper channel of the river. He had much difficulty in meeting with his old friend the king, who had been driven from the opposite bank of the river by the Kacunda people, who possessed the whole western bank of the river. He informed us that the fair was attended by more than six thousand people; that it was held on a sandbank; and that horses, tobes, goats, sheep, rice, &c. were exposed for sale. From the canoes going down we obtained about one hundred-weight of rice for a few gun-flints. All the large trading canoes had a pony on board, about thirteen hands high : they had little blood, and were in a poor condition.

We remained supplying the vessels with wood on the 12th. In the course of their work the Kroomen found a hive of bees. They immediately smoked out the bees with tobacco, and obtained about as much honey as would fill two buckets. It was the first we had found, and, though not equal to the heather, was very acceptable : it had a strong, harsh, aromatic flavour. Dr. Briggs procured a fine specimen of rock from the eastern bank, which appeared to be exclusively composed of mica slate.

Having laid in a stock of fuel, and Mr. Lander having found the passage to Bocqua intricate and shallow, steam was got up in the Alburkah to lead the way. We contrived to make an engineer out of the steward of the Alburkah, and it was expected that he would perform the duty of one as well as men of greater pretensions. Mr. Hector, the purser, was the only one of the Quorra's crew able to stand, and I was very weak and excessively nervous. To add to my nervousness, we touched once or twice ; but we scrambled on until eleven A. M. when we were obliged to anchor, in consequence of our young engineer having opened the connexion between the boiler and the deck-hose, instead of the sea-cock. The pressure of the steam blew the scalding water over the decks, and the whole of my black crew began capering about like madmen, as the water scalded their feet. I had great difficulty to prevent them from jumping overboard, and still

greater to get them to draw the fires. At the
expense of scalding my feet and hands a little,
I got down to the safety-valve and blew off the
steam, and by that means saved the boiler, as
the water was nearly blown out, and the pressure
would have destroyed the flues. The exertion
and excitement were too much for me, and I felt
that I was about to have a relapse. I sent the
boat to Mr. Lander, who determined to proceed
to the junction of the Shary and Niger, and,
should he find water sufficient, to return for us.
Having taken some goods on board the Alburkah,
he proceeded up the river in her, leaving me at
anchor. I had a smart return of fever, with intense pain in the head and other symptoms
which gave me fair warning of what I had to expect; and it was distressing to me to see Hector
in the same state.

By the morning of the 24th, I was again
better, and for the three preceding days had
been rapidly recovering from my last attack;
but I had again been reduced by it, and Dr.
Briggs told me he had given up hopes of my
recovery, particularly as I had refused to take
any medicine. I joked the Doctor a little about
doing without his assistance. During this at-

tack my spirits for the first time gave way: fortunately the fit only lasted a few hours, and was subdued with fifty drops of laudanum which the Doctor had given me to induce sleep, the want of which for several nights had harassed and weakened my mind considerably. Poor Hector's mind, I was sorry to see, was completely gone, being as weak as his body.

Mr. Lander came down from the Alburkah on the 20th, which vessel he had left at the junction of the Shary and Niger. He stated that the navigation was very intricate, but that in the true channel there was abundance of water. The inhabitants of all the towns that he had passed were delighted, by his account, at the prospect of trading with white people, and the country he described as being even more beautiful than that where we then were. On the 21st we got under weigh and passed Bocqua, the old chief sending us a boat-load of wood, and entered the gorge of the Kong Mountains. The river here is confined in a bed of about seven hundred yards wide, the Kong Mountains rising to the height of from two to three thousand feet on each side. On the western bank there is a mountain which bears so strong a resemblance to Arthur's Seat near Edinburgh, that Dr. Briggs and myself were struck with the resemblance at the same instant. On the eastern bank there are three sides or lips of what appeared to me the remains of an immense crater. The side facing the north-west, which was that next to the river, had fallen, and immediately before it stood one of the most remarkably shaped sugar-loaf hills I ever saw. It was a perfect cone about five hundred feet high, surmounted by a curious column of rock, which, on examining it through a telescope, Dr. Briggs pronounced to be basalt. It certainly had very much the appearance of it, but the distance may have deceived us.

The Kong Mountains lie in the direction of about west-north-west and east-south-east, where they are intersected by the Niger: they have a remarkably bold outline, and, if we might judge by those portions of them that abutted on the river, appear to be composed principally of granite. The channel between them is dangerous, owing to large blocks of granite that lie in the stream, producing eddies and shoals. On one of these we ran aground about four in the afternoon, but got off again in about half an hour, and came to anchor for the night below the rocky islets that appear to guard the upper entrance of the strait. In the morning we were again under weigh, and a few minutes afterwards opened one of the noblest reaches that imagination could have conceived. An immense river, about three thousand yards wide, extending as far as the eye could reach, lay before us, flowing majestically between its banks, which rose gradually to a considerable height and were studded with clumps of trees and brushwood, giving them the appearance of a gentleman's park ; while the smoke rising from different towns on its banks, and the number of canoes floating on its bosom, gave it an aspect of security and peace far beyond any African scene I had yet witnessed. The confluence of the Shary was just in sight, and a range of low hills on the northern bank trended east-north-east ; while on the western bank of the Niger were two remarkable isolated table-lands of a romantic and beautiful appearance, giving a finish to a picture to which no description can do adequate justice.

At ten A. M. as we were pursuing our course, the Quorra struck on a shelf of rocks extend-

ing from the bank to a considerable distance. I
thought the shock would have been fatal to the
vessel ; but she fell off of her own accord, and,
finding that she made no water, we proceeded
on. Shortly afterwards we ran on a sandbank
in attempting to cross the river, and passed the
whole day in vainly trying to get off again.
We blew off the boilers to lighten the vessel,
but to no purpose. Mr. Lander left us and went
up to the Alburkah, which was in sight, about
six miles distant.

In the morning of the 23rd, we sent the carriage-guns, round and canister shot, on shore,—
landed the best bower cable, and all the weighty
things about the decks : about two P.M. the vessel came off, after about twenty-four hours' hard
labour.

Mr. Lander returned from the Alburkah, and
informed me that Lieutenant Allen had a severe
relapse, and that Captain Hill was better ; but
that the carpenter, a fine young man, was
dead, being the third man they had lost in
the river.

Christmas Day found most of us gradually
recovering. I was gaining strength rapidly, and
so was Dr. Briggs. Hector was also better ; but

John, the second steward, was worse, and I was apprehensive that we should lose him. In the afternoon we made another attempt to get the vessel across the river, having sounded all the way in a boat. We had lightened her to four feet seven inches, but had scarcely got under weigh when she grounded again, running her bows a foot out of the water. Mr. Lander was quite disheartened and very unwell. He insisted on returning to the Alburkah against the advice of Dr. Briggs, who thought him predisposed to dysentery, and was afraid the exposure would bring it on.

The vessel being only aground under her bows, we got an anchor out astern, and prepared to discharge from the fore-hold in the morning. The thermometer this morning was at 66°, and at three P. M. it was at 90°, the warmest Christmas by far of any I had passed before,—but a fine breeze tempered the afternoon heat. Dr. Briggs and I passed, I will not say a merry Christmas, but a thankful one, and toasted our friends at home in

" Cups that cheer, but not inebriate."

CHAPTER VI.

*Sickness in the Alburkah.—Disappointment in the Navigation.
—More Sickness in the Quorra.—Prepare for the Rainy
Season.—Another Messenger from Attah.—Attah Canoes.—
Attempt of Mr. Lander to proceed to Boussa.—A Thief
caught.—Visit to Addakudda.—Description of the River
above Eboe.—Superior Character of the Natives.—Manufactures and Trade.*

At daybreak on the following morning we
discharged two boat-loads of goods from the fore-
hold and hove the vessel off. In attempting to
warp her across the river, she grounded again,
after getting half way over, in consequence of the
kedges coming home. It became necessary to
lighten her more. Fortunately she had grounded
on sand, and remained perfectly tight.

At daybreak on the 27th, we found that a bank
had formed under her larboard quarter, on which
there was a depth only of two feet water, while
on the opposite quarter we had ten feet. The

two kedges were now carried out astern, and we hove her off about ten feet clear of the bank, when the kedges came home, and she drifted on again in a worse position than before. We set to work and discharged more cargo, by which we lightened her to four and a half feet on an even keel. In the evening I received a note from Mr. Lander, informing me that he was seized with intermittent fever, and that he despaired of Lieut. Allen, whose head he said was much affected, and that Captain Hill was confined to his bed.

On the 28th, I was feverish and unwell all day. The men were employed using every means to get the vessel off, but in vain. The sands kept shifting, and no anchor would hold in them. This morning the vessel was aground, head and stern, the current bubbling and roaring beneath her for about twenty feet amidships, where there was twelve feet water. She now appeared to be hard and fast amidships, the current running like a sluice under her fore and after body: indeed, it was some matter of surprise to me that she was not broken fairly in two.

In the evening I felt very unwell, and considered that I was about to have a relapse. Dr. Briggs was in the same state. I took an emetic and six grains of calomel, and resigned myself to my fate.

In the interval between the foregoing and the 7th of January, Dr. Briggs and I experienced another severe attack of fever, and it was not until this date that I could resume my journal. We were both like scarecrows, with our long beards and razor faces. On this day, however, I was much better, and Dr. Briggs was convalescent. The vessel still lay aground. Such detention and delay, and no trade withal, was very distressing. We had now been upwards of two months in the river, and had not obtained half a ton of ivory. However, we had not yet reached the places where Mr. Lander saw it in such profusion; but after all, when we were asked at the rate of one shilling per pound for it, I was very doubtful of its cheapness. The indigo was four-fifths of it dirt, and would not pay freight home.

On the 8th, I felt nearly well. The vessel could not be got off. The men had been heaving on the kedges in vain, and I began to think that if she were afloat, I should be rather at a loss what to do with her; for I doubted whether we could remove her from the hole which she had made for herself, surrounded as she was by banks with three or four feet water only on them. On the previous day I had purchased a bullock for four yards of cloth; and after enjoying a beef-steak, Briggs and I found ourselves equal to a little walk on shore. On landing we visited a small village on the eastern bank of the river, and took a guide to the top of the bank, which was about one hundred and fifty feet high. From the summit we had a magnificent view. The country appeared thinly inhabited towards the interior, compared with that near the banks of the river, and was more thickly wooded.

Dr. Briggs collected a quantity of minerals in our walk, and we met with the wild cotton, tamarind, and a number of beautiful shrubs. The bank was formed of granite, covered over with a stratum of rich vegetable mould, with here and there large boulders of granite, of fantastic shapes, rearing their naked heads above the long grass. We soon after returned on board much benefited by our little excursion. My friend Briggs was an enthusiast in geology, and was hammering at every stone we met with. Among other acquirements which he had picked up in his travels on the Continent was that of cooking;

—and let no traveller sneer at the accomplishment.

In our own case, as we had been without such an assistant for a considerable time, and had only a Krooman to cook for us, his abilities in this line had been and were frequently in requisition for our mutual comfort. Whenever Mr. Lander visited us, we were better off, as he always brought his own cook, Pascoe, with him, who might have studied under Ude himself. As both our stewards were gone, we had promoted two boys we found at York to their office, and they had been our only attendants during sickness. These two were both Eboes, and spoke the language very well. My valet's name was Friday, and the Doctor's Saturday. Friday was about fourteen, and his history was somewhat curious. He was once sold in the Bonny to a slaver, and made a voyage to Havannah. Being a sharp intelligent lad, the captain kept him as a cabin-boy; and on the next voyage to the coast the vessel was captured, and Friday was landed at Sierra Leone, and from thence sent to the Banana Islands. He there gave himself out as being the son of the King of Eboe; and our hospitable friend Mr. Pratt of

York, thinking our expedition was a favourable opportunity for his return, let us have him. It may be needless to add, that the story of his royal descent was his own invention. But he proved a useful and apt servant, and I intended, if he continued so, to bring him to England.

On the 9th, Mr. Lander came down to us from the Alburkah, looking much reduced by his attack of intermittent fever. He appeared in very low spirits about our prospects, which certainly were bad enough. He had been up to within six miles of Cuttum-Curaffee, which town he reports as being that distance from the river. It has lately been burnt and plundered by the Felatahs, and, in consequence, nothing was to be had there. It was expected by the natives that the rains would commence in the following month, and he had begun to house in the Alburkah, recommending me to do the same with the Quorra. It was a measure absolutely necessary with us, for our decks were like sieves, and how we were to caulk them was beyond my comprehension, as not one of my men was able to do any work without great risk of a relapse.

A messenger from our old acquaintance of Attah came up to-day in a large canoe, bringing

as his credentials a porter-bottle, which he displayed with great gravity and importance, uttering at the same time the magic monosyllable
'rum.'   He informed me that his master had
collected four hundred teeth, and wished us to
return and buy them.   Mr. Lander proposed to
go down in the launch; to which I strongly objected, as it was evidently a *ruse* of the old rascal
to get a white man in his power.   The visit of
this emissary terminated in our filling his bottle
with rum as a sign that he had been with us,
and we despatched him to his master with a
message, saying we would send old Pascoe down.
Mr. Lander returned to the Alburkah in the
evening.

In the afternoon of the 10th, I went to the
Alburkah, and arrived at midnight completely
exhausted, having been five hours in going as
many miles.

On the following day I went on shore, with
Mr. Lander, to examine a situation preparing for
a house to contain our goods.   It was on a small
hill about two hundred and fifty feet above the
river, that commands the junction of the Shary
and Niger.   The view from it is very fine, and
I do not think that a more desirable spot could

be found. The steamers could lie close alongside the bank; the ground was dry; and if we succeeded in erecting any building (which I then
very much doubted), it appeared there was every
chance of its being healthy. I returned to the
Quorra in the evening, and found that she had
been got off, and, as usual, before she was ten
yards from the place where she lay, had run
aground again.

On the morning of the 12th, we made a last
effort to heave her off again into deep water, but
in vain; and as she remained perfectly upright, I
determined on the next day to re-ship the cargo,
and house her over for the rains. We had been
eighteen days afloat and aground, frequently in
situations of great danger, and the men were
completely worn out. In the evening I experienced another return of fever.

The next day being Sunday, the 13th, Dr.
Briggs read prayers. A canoe came down from
the Alburkah with a letter from Mr. Lander, giving me a fearful account of the danger to which
he had been exposed from some large Attah
canoes. Six of them, it appeared, had been paddling round the Alburkah all night, but did not
venture to board. This King of Attah, it seemed,

was the ruling man in this part of the country: Fundah pays him tribute, and all the towns on the banks of the river are kept in awe by his numerous canoes, which are continually lurking about stealing children and attacking the defenceless villages. Scarcely a night passed but we heard the screams of some unfortunate beings that were carried off into slavery by these villanous depredators. The King of Attah's territory extends to the Shary, but how far to the eastward we had no means of judging. The messenger that he had sent up, to say that he had a quantity of ivory, was an impostor; and the whole, as I suspected, was an artifice to get Mr. Lander or myself into his power. When they found in the canoe that we were aground, they proceeded up the river, and calling at every village, denounced their master's vengeance on all that traded with us: the consequence of which was, that we could buy nothing alongside, and were entirely dependent on the mountaineers inhabiting the hills under which the Alburkah lay. These people, in the true spirit of their race, acknowledged no ruler but their own.

I have already alluded to the remarkable table-lands which we saw, but had no idea of their extent.

On that nearest to us were four large towns, rich in sheep, goats, and bullocks. The paths leading up to these table-lands are so steep, that Pascoe, who was despatched on a mission to them, was carried over the most difficult parts, being quite unable to get over them otherwise. Secure in these mountain fastnesses, they laugh to scorn the kings of the plains below, with whom they are generally at war. I regretted much that the state of my health did not permit me to visit them.

In the course of the day I had a severe attack of ague, which I anticipated would be my daily companion.

At daybreak on the 14th, I sent an armed boat with Harvey and Kirby (the only two white men on duty), and six Kroomen, to the Alburkah. In the afternoon two large canoes came alongside, from which I purchased some rice and yams. They belonged to Attah, and were of the party which had produced so much alarm, with so little cause, on the mind of Mr. Lander. They had been at war with a town above Cuttum-Curaffee, which they had destroyed, and had taken a number of slaves, some of whom they had on board, and were anxious to dispose of to us. I suppose it was nothing more nor less than a slaving expedition on a larger scale than usual. They complained bitterly of Mr. Lander, for not allowing them to go alongside of his vessel, and for threatening to fire into them, who, they added, were his best friends. By way of amusing them, I directed a twenty-four pounder to be loaded, and discharged a shot along the surface of the water; and their delight on seeing it strike and *ricochet* to the opposite bank was equal to their surprise. At their earnest request, I sent a shot by them to their king, as a specimen of the kind of bullets which we were in the habit of using.

In the evening our boat returned with a letter from Mr. Lander, informing me that the war palaver was over; but his imagined enemies had already told me that it had never existed on their side at all events. Both Dr. Briggs and myself were suffering under a violent attack of ague.

On the 15th, we re-shipped the greater part of our cargo. As a proof of the security of trade in that part of Africa, I may here mention that the goods had been lying on shore for three weeks, and we had not lost a single package of any description ; their only guard the whole time being two Kroomen, who spent their time more in shooting monkeys than guarding the goods. Fever and ague were my visiters, as usual, in the afternoon.

At daylight on the 16th, I sent a boat to the Alburkah with goods ; but I found myself much worse, and indeed not able to stand from excessive debility. Dr. Briggs was better, and appeared to have got rid of his ague. On the return of the boat, I was sorry to find that my friends were again in hot water—could get no provisions, could buy no ivory, and in fact could do nothing : but, to my great relief, they spoke of moving up to Rabbah, and if possible to Boussa. I was rejoiced at this prospect on more accounts than one ;—the question would thus be settled as to the existence of the quantity of ivory we had been led to expect, but of which we had as yet seen nothing ; and it would enable me to find subsistence for my own crew, as ever since our arrival we had been living from hand to mouth, seldom having two days' fresh provisions on board, and frequently were obliged to encroach on our salt provisions, which I was carefully hoarding for any case of emergency.

On the 17th, I remained very unwell, and
began to fear that I was becoming rather hy-
pochondriacal, which in the climate we were in
was likely to be productive of bad results.   I
sent a note to Mr. Lander, informing him of
my inability to move, and my anxiety that he
should proceed without delay to Rabbah and
Boussa.

Mr. Lander and Captain Hill came down to
the Quorra on the following day.   The former
appeared better, and the latter I thought was
very well.   They informed me, that men employed by the King of Attah were stationed in
every market-town, to prevent the natives from
trading with us for provisions or ivory; and Mr.
Lander considered it the best plan to proceed
at once up the river.   I quite agreed in the pro-
posal, and recommended him to take my largest
boat and six or seven Kroomen with him, in case
of meeting with any hostility; and it would relieve me of so many mouths, which, in the present state of the provision market, was of some
consequence.   This however he declined, saying
he had more men than he knew what to do
with, and disliked Kroomen.

Eventually it was arranged that the Alburkah

should proceed with goods suitable for the upper
country, and that I should follow in the Quorra
as soon as she floated, which we expected would
be in March. After a short stay on board, Mr.
Lander and Captain Hill returned to their vessel, leaving a list of goods which they wished
to be sent after them. I wished very much that
they would go, as they were constantly having
some misunderstanding with the natives, which
I attributed to Mr. Lander's paying too much
attention to the stories that his own men Pascoe, Jowdie, and Mina were always bringing
him. I have ever found the natives peaceable
and well-disposed, and, with the exception of
their pilfering propensities, easily managed.
Stealing I have never overlooked, and, during
the time we were aground, I flogged above a
score of them for theft, both masters and slaves.
One fellow was caught so neatly in the very
act, that I cannot forbear relating the circumstance. Dr. Briggs was lying on one of the
sofas in the cabin, reading, when a woolly head
made its appearance at one of the side ports,
and a black paw seized upon my dressing-box,
which was lying within reach from the window.
The Doctor made a spring, caught the fellow by

the ears, and called out lustily for assistance. As the fellow's companions in the canoe were hauling him down by his legs, one of the men jumped down to the cabin to relieve the Doctor, and two of my Kroomen stepped into the canoe and gave the culprit thus caught *in flagrante delicto* three dozen lashes in a style that would not have disgraced any boatswain's mate in his Majesty's service. I never found that this severity did any harm; in fact, it was always applauded by the natives, who, thieves themselves, think in common with their more civilised brethren, that the disgrace lies not in the act, but in the detection of the theft.

I forwarded a large quantity of our most valuable goods to Mr. Lander for his intended voyage to Rabbah and Boussa, and received in return the ivory which he had collected, being something less than three hundred weight, as the result of our trade for two months.

Accompanied by Dr. Briggs, I paid a visit to Addakudda, the largest town in sight from the vessel on the western bank of the river. It is prettily situated on huge blocks of granite, forming a natural barrier on the side next to the river, and giving it the appearance of a fortified

place. Betwixt two of the blocks, is a landing-
place which might have been formed by art.
The town contains about five thousand inhabitants, and, like all the African towns which we
have yet met with, is abominably dirty and irregularly constructed. The chief, to whom we took
the usual "dash," received us very graciously,
and presenting us with goora nuts, chatted with
us very cordially for some time. I told him that
I had opened a market on a sandbank near my
vessel, as I had heard his people did not like to
trade direct with white men. He assured me of
his friendship, and said he was very glad that I
had punished his men when they were detected
stealing, particularly his son, whom he considered
a great rascal, and who richly deserved it. I
was not a little amused at finding thus unexpectedly that it was the heir-apparent who had received the chastisement above alluded to. However, the chief was polite enough to show us over
a large dyeing-ground, which was placed on an
artificial mound of earth raised about thirty feet
above the level of the village, and covered with
poles and sticks for drying the cloth on as it was
taken from the pit. Their process appeared to
be exceedingly rude, and was briefly as follows:

A quantity of crude indigo, made up in balls of about three inches diameter, is thrown into a pit, about four feet wide, and eight feet deep. This pit is filled with water, and a ley of wood ashes being added, is allowed to ferment : when the fermentation has subsided, a board is put down into the pit, the sediment is pushed on one side, and the blue liquor remains. The pit has a small gallows erected over it, and the piece of cloth to be dyed is suspended from it by a string, and let down and hauled up by the dyer until he considers the dye to be sufficiently deep.

The beauty of their dye consists, I think, in the freshness of the indigo, and the quantity they use. The price of dyeing a tobe, which if well made contains from eleven to fourteen square yards, is two thousand cowries, or two shillings sterling. There were about fifteen or sixteen of these pits on the hill, each with a separate proprietor.

It being market-day, the town altogether presented a lively scene of trade and industry. In the market we found palm-oil, shea butter, four or five kinds of grain, Cayenne pepper, calavances, yams, &c. exposed in considerable quantities for sale. The women who sold them were loud in the praises of their goods, and emulous of the patronage of the white men, who, I imagine, are made to pay about four times the market-price for everything which they buy.

As usual after any exertion, we had a much more violent attack of the intermittent on our return than usual. The daily return of the ague had weakened me, and reduced me so much, that I feared if it lasted much longer my mind would suffer. Dr. Briggs was much better, and had only occasional attacks; but I had scarcely escaped a day for the last three weeks. Harvey was the only seaman on duty, and he had become a scarecrow. As for the rest, they were sunk into a deplorable state of despondency. But this was not to be wondered at, as the poor fellows had no kind of resources within themselves, and the *ennui* of their situation, combined with the lassitude produced by the climate, gave rise to anything but a pleasant state of mind.

I had been busily employed during the last few days in building a hut on an island distant about half a mile from the vessel, which I intended to open as a shop on a sort of debateable ground, as the natives did not appear very ready to bring their goods on board the vessel. In clearing the ground for the hut we started two wild swans, and Dr. Briggs winged one of them with his fowling-piece, but he was too expert a diver to be taken. We soon after found their nest with nine noble eggs in it, which the Doctor speedily converted into a savoury omelet.

On the 29th of January we opened our market with a salute of nine guns, from two four-pounders that I had landed there, having previously sent word to all the towns about us of our intention. Few people attended, and we only took ten thousand cowries on our first day. Dr. Briggs superintended, as I was confined by a severe attack of fever, and suffered for his kindness, as the exertion brought on a smart return of intermittent.

As I passed many happy as well as melancholy days at this place, I will here introduce a general account of the country between Eboe and the confluence of the Shary and Niger, some description of its inhabitants, their manners and customs, its productions and physical formation, with the mode of life which we led on board the Quorra, as my daily journal would be uninteresting on account of its sameness, and useless from its repetition.

On leaving Eboe we emerged from a comparatively winding and narrow stream bordered by stagnant swamps and overgrown with immense forests, the sameness of which distressed the eye, while the extent baffled the imagination, into a wide and splendid river. The banks were but thinly wooded, and in many places highly and extensively cultivated. The various reaches of the river became longer, and in its serpentine course it assumed a more graceful character, while the inhabitants on its banks were more civilised and better apparelled. We found the better class attired in the Houssa loose shirt and trousers, instead of the common wrapper of the Eboes. The country generally presented that formed and decided appearance which characterises land that has been long under the dominion of man. The banks, although elevated fifteen or twenty feet above the surface of the river, continued flat until we arrived at Kirree, where we met with the first bluff. From thence the country gradually rises, until at Attah it attains an elevation of from two to three hundred feet. From Attah upwards a range of hills on each bank of the river gives the scenery a picturesque and bold character ; those on the western bank seem to have the highest elevation, but neither appeared to rise above four hundred and fifty or five hundred feet above the water. From the general outline of these hills, and from the specimens of the rock that we found being principally granite and mica slate, we pronounced them of primitive formation.

Passing through this romantic valley, which extends from forty to fifty miles, we reached the Kong Mountains, which on the banks of the river rise to an elevation of between two and three thousand feet. As far as we could ascertain, they are composed principally of granite, and have a bold and magnificent appearance. The chasm through which the river passes seemed about fifteen hundred yards wide, but the channel of the river does not occupy more than seven hundred.

It would not become me, with the very superficial knowledge of geology that I possess, to venture on an opinion of the convulsion of nature that opened the passage through these mountains. That it must have been beyond all human conception, is evident from the number of huge masses of granite of the most fantastic shapes projecting out in the most singular positions in all directions. From the narrowness of the gorge and the abrupt rise of the mountains from the river, if future travellers confirm the opinion of Dr. Briggs and myself of the existence of the remains of an immense crater on the eastern side of it, there may be grounds for supposing that volcanic agency has been the principal means employed; but as neither of us were able to land there, our opinions must be considered as conjectural.

After passing the Kong Mountains, a beautiful reach of the river, about fifteen miles long, with an average width of three thousand yards, stretches to the confluence of the Shary, where there are several rocky islands. One of these presents a most curious appearance, which I can compare to nothing else than to that part of Arthur's Seat called " Samson's Ribs," provided they were laid horizontally and twisted in the most fantastic curves. I observed that the bed of the river was changed in its character after leaving the alluvial soil. There were no mud banks: they were now formed of coarse sand, which was continually shifting, — so much so, that an anchor in a few days always became embedded, and we never could get a kedge to hold in our repeated attempts to get the vessel off. Another peculiarity of these banks is their abruptness : frequently I have had four fathoms within a foot of a bank on which there were not as many feet of water, the current appearing to cut it almost perpendicularly.

Both banks of the river are thickly studded with towns and villages. I could count seven from the place where we lay aground; and between Eboe and the confluence of the rivers, there cannot be less than forty, generally occurring every two or three miles. The principal towns are Attah and Addakudda; and averaging the inhabitants at one thousand for each town and village, will, I think, very nearly give the population of the banks. It may be rather below the mark.

The general character of the people is much superior to that of the inhabitants of the swampy country between them and the coast. They are shrewd, intelligent and quick in their perception, milder in their dispositions, and more peaceable in their habits. The security of life and property is evidently greater among them ; though it is still sufficiently precarious to prevent the inhabitants from living in isolated situations, nor will any of them venture upon the river after sunset in small canoes. Agriculture is extensively followed, and Indian corn and other grain are raised with little labour and less skill on the part of the cultivators. Tobacco is grown sparingly, and when dried and made up for sale, costs one hundred cowries, or one penny per pound. It has a mild, pleasant flavour, and is made up in rolls in the Turkish fashion. The natives are greatly addicted to smoking, and use the long reed pipe common in the Levant. The bowls of these pipes are neatly manufactured of clay : a few copper pipes are met with, that are brought from Fundah and other towns on the Shary.

Beer or sweetwort is manufactured in large quantities from Indian corn and other grain, and markets for the exclusive sale of it are held periodically.* That made from dhoura is a pleasant, agreeable drink, but apt to produce diarrhœa. — Yams, calavances, &c. are plenti-

---

* The stillness of the night was frequently broken by the passing of canoes full of *bons vivants* returning from these markets.

ful, the former much inferior to those of Eboe.
The river abounds in fish to a degree that is
almost inconceivable, and the inhabitants of
the banks are expert and persevering fishermen. They make immense nets of grass, which
they use as seines with great dexterity. They
are very careful of their nets after using them,
and stretch them on poles to be dried by the sun
exactly as our fishermen do. The fish are split
by them and gutted, they are then dried by the
smoke of a wood fire, and form with farinaceous
food their principal means of subsistence.—Fruits
are not plentiful on the banks of the river : plantains, bananas, limes, tamarinds, a species of
plum, and pine-apples, constitute the whole. The
latter are exceedingly scarce, and the former by
no means abundant.

The intercourse and trade between the towns
on the banks is very great, (I was surprised to
learn from Dr. Briggs that there appeared to be
twice as much traffic going forward here as in
the upper parts of the Rhine,) the whole population on the Niger being eminently of a commercial character, men, women, and children
carrying on trade. The traffic in slaves, cloth,
and ivory is confined to the men ; everything else

being left to the other sex, who, to say the truth,
are far the most difficult to deal with.

Bocqua, or Hickory, as the natives call it, is
the centre of this traffic; and a fair of three days'
duration is held there every ten days, attended by
Eboe and Attah, and even Bonny traders from the
south, and those from Egga, Cuttum-Curaffee,
and Fundah on the north, besides great numbers
from the interior country on both banks of the
river.   The traders from the upper country
bring cloths of native manufacture, beads, ivory,
rice, straw-hats, and slaves, all of which they sell
for cowries, and buy European goods, chiefly
Portuguese and Spanish.   About twenty-five
large canoes passed us every ten days, on their
way to this market, each containing from forty
to sixty people.   The trade is carried on by
money, not by barter: cowries are the circu-
lating medium, and their sterling value on an
average may be taken at one shilling per thousand.   The cowries are strung together in portions of one and two hundreds, each portion being on a separate string; but it is necessary to
count them, as the natives always endeavour to
cheat.   The following extract from my trade

journal, will show in a clearer view the barter trade :—

" Two scrivelloes, weighing eleven and a half pounds, were bought for nine cotton handkerchiefs and one cutlass, or about 4s. 4d. sterling.

" Two scrivelloes, weighing eighteen pounds, for two yards scarlet cloth, one yard cotton velvet, and four handkerchiefs; about 5s. 6d. sterling.

" One tooth weighing thirty pounds,—three yards red cloth, and eleven thousand cowries, or about 14s. sterling.

" One tooth weighing forty-four pounds,—four yards of red velvet, six yards of scarlet cloth; about 16s. sterling.

" One tooth weighing thirty pounds, — four yards scarlet cloth, one looking-glass; about 7s. sterling."

Hardware, powder, guns, Manchester cottons, and earthenware were not saleable at prime cost: red cloth, velvet, and mock coral beads were principally in demand; and next, looking-glasses and snuff-boxes of Venetian manufacture. I have stated the value of the goods at the invoice price; but the price paid for ivory is no criterion of the value of the trade, as the quantity is so small that the expense of collecting it will prevent the success of any commercial enterprise dependent on that article alone. I have been a month without having a single tooth offered me for sale, and I believe the natives never passed us when they had any to dispose of.

The trade is tedious and excessively trying to the temper. A tooth is generally owned by four or five individuals; and as each acts independently of his partners, it is generally brought on board as many times before its actual sale. I found the best plan was to offer them at once one half of what they asked, and my offer was generally accepted. The traders from the upper country usually employed a broker named Mallam Cantab, an inhabitant of Addakudda, to trade for them, and paid him a regular commission: in fact, the old rascal got paid on both sides; as I did not discover for some time that this was the common practice of the country. The best and most perfect ivory comes from the Shary; the teeth from the Niger are almost invariably broken at the end.

I have previously remarked that the indigo is four-fifths dirt, and cannot in its present state be considered as an article of commerce, although eventually it may become one of the staple products of the country. I never met with bees' wax in cakes; but, from the abundance of honey, it is not unlikely, if the natives knew its real value, that it might be collected in considerable quantities. A few ostrich-feathers were offered me; but they were badly preserved and of a poor quality. Leopard-skins were frequently offered us, but not in sufficient quantities to be considered as an article of trade: their price was from two thousand five hundred to three thousand cowries.

The price of provisions varies considerably: we purchased them principally from canoes, the owners of which, I suppose, like bumboat women, charged well for their trouble in bringing them alongside. Bullocks cost from twelve to twenty thousand cowries each, and weighed from sixty to one hundred and twenty pounds; goats were from one thousand to one thousand five hundred cowries, weighing from eight to fifteen pounds. Fish we found very cheap: a handkerchief would buy sufficient for a day's consumption. Rice is bought in small grass bags, which hold about two pounds: the price generally was one gun-flint per bag;—it is sweet, but badly cleaned;

yams are scarce and dear, and of a bad quality. Fowls are plentiful at one hundred cowries apiece; and eggs are abundant, such as they are, nineteen out of twenty being generally rotten.

CHAPTER VII.

Amusement on Board.—Reflections on the State of the Inhabit-
ants.—Native Canoes.—Mr. Lander unsuccessful.—Change
in the Season.—Further Sickness.—Death of Dr. Briggs.—
Tribute to his Memory.—Departure for Fundah, the Quorra
left aground.—Arrival at Yimmahah.—Reports of the King
of Fundah.— Friendly Mallam.—Mode of thatching Huts.
— The Native Barber.

THE life we led on board was monotonous in
the extreme, and if Dr. Briggs and myself had
not been well provided with books, would have
been more so. Whenever we were able to do it,
we were up at sunrise, and breakfasted at eight;
at one we dined, and were generally in bed at
eight in the evening. Sleeping, as we always
did, on deck, we had frequent opportunities, after
retiring to rest, of contemplating the beautiful
scene which the river presented, as its noble
waters rolled silently and majestically by us, when
lighted by a moon such as those only who have
been in the tropics can imagine. After a distressing and enervating day, the approach of night

was always hailed by us with delight: its calmness and tranquillity, its silent grandeur and its
beauty, made us for a time forget our own personal sufferings, and sometimes almost deem such
enjoyment cheaply purchased by them.

My mornings were generally passed in reading;
while Dr. Briggs studied Arabic. A game or two
at chess helped to pass the afternoon away, when
an attack of intermittent did not disable us from
attending to anything but our own miserable
bodies. The attacks of ague generally came on
at noon, and each successive one was invariably
half an hour later, until it arrived at six o'clock,
when it would leave me for two or three days
and then begin again at noon. This order was
repeated so often in my case, that I could tell to
a minute when it was coming on.

Shakspeare and Scott were my favourite authors, and I can bear my humble testimony to
the value of the works of those great magicians
to such as " are sick and afar off." Many an
hour that would otherwise have been spent in
useless repining or melancholy foreboding, I
have passed not merely placidly, but happily,
in their perusal; forgetting the present, living
in the past, and heedless of the future.

Before leaving home I had purchased Blackwood's Magazine from its commencement, and was very much struck one day when looking over it by a letter of Mr. M'Queen's to Lord Goderich, in which he recommended that the Benin, or some of the numerous rivers that flow into the sea between that river and the Calabar, should be explored; stating his reasons for supposing that the Niger emptied itself into the sea by these mouths. His opinion has been verified by experience; and discovery has so fully proved the accuracy of his descriptions, that it is difficult to imagine that he must have been anywhere than on the spot when he penned that memorable letter, which will ever remain an evidence of his sound judgment and geographical skill.*

Amongst our other employments, I should not omit to mention our culinary labours, which were in daily request. The meat which we obtained was generally goat's-flesh, and as Dr. Briggs and I agreed in taste, we claimed the head and trotters for the cabin table, and either

---

* It may be truly said that the honour of the theory belonged to Reichard; but his opinion was founded on the extent of alluvial deposit alone: it was strengthened and confirmed by the additional evidence collected with so much zeal and ability by Mr. M'Queen.

stewed or made them into a broth. Roast or
baked meat we found heavy and indigestible,
and we invariably stewed or boiled everything.
Coffee was our only beverage, of which we partook in large quantities during the day.   On the
whole, we had reason to be thankful that we had
always abundance of the necessaries, as well as
not a few of the luxuries of life ; and on comparing our condition with that of Park and other
travellers in the interior of Africa, we had ample
reason to congratulate ourselves on having a
roof to shelter us, and a bed on which we could
repose.

Since we had left Eboe, I had never any watch
kept at night, as I considered that we were
perfectly secure among these harmless and amiable people.   I speak now of the character of
the lower and middle classes of them generally:
there are many exceptions to this character
among the chiefs ; but the lower classes are kind,
hospitable, and obliging.   It is true they have
the faults and evil propensities of the negro,
being liars, cowards, and thieves : but they have
these qualities engrafted on a naturally good
disposition by a trade that would demoralise the
Scotch and make cowards of the English ; and

it is some matter of wonder that the centuries in which the slave-trade has been in existence among them should have left them any good qualities whatever.

In paddling their canoes, the natives have a pleasing custom of singing; and the distant sound of their voices, as it falls on the ear, produces a very soothing effect.   One of the canoemen, who is generally the steersman, and always standing, leads the song; and the whole crew join in chorus, keeping time with their paddles to the measure.   On nearing our vessel, they generally made us the subject of these effusions, and never omitted standing up in the canoe, and saluting us with a good-b'ye or a good-night in the Houssa language as they passed, being the language generally spoken by traders.   The distance which they get over in these large canoes is astonishing : they think nothing of paddling for twenty-four hours without ceasing, except to have refreshment for a very little while ; and by taking advantage of the eddies of the river, with which they are well acquainted, they will average against the stream five miles an hour.

In the afternoon of the 5th February, I was much surprised and annoyed on receiving a

letter from Mr. Lander, informing me that there
was not sufficient depth of water for the Alburkah, and therefore that she could not proceed
to Rabbah and Boussa, as was intended. It
stated that Captain Hill had sounded the river,
and that he could walk across it.

For the last fortnight the weather had been
sultry and close : fortunately we had roofed the
Quorra in with long grass, and prepared for the
rainy season, which was fast approaching. We
had had two or three squalls, but no regular tornadoes. The thermometer had varied from 82°
to 97° between seven in the morning and four in
the afternoon. The evenings now were no longer
pleasant and cool, and we found them even more
distressing than the days. Sickness had done its
work on us all. My boatswain, Harvey, was the
only man who had tolerable health ; the remaining three seamen were confined constantly to
their hammocks. I was emaciated to the last
degree, while my joints were swelled to twice
their natural size, and excessive debility disabled
me from walking or even standing without assistance. Dr. Briggs, although not so much reduced as I was, had one or two attacks of
dysentery, which alarmed me exceedingly. He

had been much on shore, being very anxious to learn the Houssa language, and to become familiarised with the habits and manners of the natives, and I was apprehensive that the exertion he had used would injure him.

Mr. Lander now proposed to send Pascoe down the river with letters to the brig, which I had no doubt he would accomplish without any personal risk : and he informs me that the King of Fundah is exceedingly anxious to see one of the white men. In consequence of this, I entertained an idea of going up the Shary, and gave directions for the largest boat of the Quorra to be fitted out, and a hurricane-house to be erected in the stern-sheets. Dr. Briggs was of opinion that the change would benefit me, and that he could take charge of the vessel till my return.

On the 8th of February we had a terrific tornado, with its usual accompaniments. However, it cleared the air, and the morning was much cooler and pleasanter than any which we had experienced for some time previously. But the sudden change of temperature was too much for my debilitated frame, and it brought on a severe attack of remittent fever, which lasted for several days and had very nearly carried me off.

On the 18th, Mr. Lander, who had been so kind as to come down to the Quorra for two or three days on hearing of my relapse, considering that I was out of immediate danger, returned to the Alburkah, having determined to accompany me to Fundah as soon as I should be equal to the exertion. I found from him that Mr. Jones, the mate of the Alburkah, had died on the 7th, after a relapse. The natives all agree that this month is the most unhealthy, and that there was now much sickness and mortality among themselves: many of the Kroomen even were complaining. The weather was intensely hot, the thermometer having been frequently at 99° for hours; and not a breath of wind tempered the heat during the day, while the nights were moist and raw without being cool.

Dr. Briggs became very ill with an attack of dysentery, and removed his cot into the cabin, thinking it cooler than on deck. Mr. Hector being pretty well, was able in some measure to attend him.

The fever had left me so exceedingly weak in mind as well as body, that after Mr. Lander's departure I lay on deck in a state of almost total unconsciousness; but I was painfully roused from my stupor by the death of my dear friend and companion Dr. Briggs.   On the 27th he was brought up from the cabin, and I was shocked to see the ravages which a few days' suffering had made on him.   He had never been much reduced by his repeated attacks of fever; but now so altered was he, that I scarcely knew him.   While shaking hands with me, he assured me, with a weak but cheerful voice, that he felt better, and forgetting his own sufferings, anxiously inquired after mine. We lay side by side for some hours, and he pressed me much to go down into the cabin, as he considered passing the evenings on deck very unhealthy; but I was incapable of moving myself, and afraid of being carried, my bones being very prominent and excessively painful when touched.   At sunset he was carried down, being then in severe pain, and I bade him farewell, little thinking it was for the last time.

On the 28th the pain suddenly left him; on which he told Hector and Sarsfield, who were attending him, that mortification had commenced, desiring them at the same time not to tell me of it till all was over.   Soon after he expired without a struggle, tranquilly yielding his spirit to Him who gave it.

I cannot describe the feeling of anguish and desolation that came over me when I was told of my bereavement. At first, I could scarcely believe the fact that my beloved friend was really gone, and gradually sunk into a state of apathy and indifference to all around, in which I continued for several weeks.

Thus died, in his twenty-eighth year, Dr. Thomas Briggs. He was the eldest son of Dr. Briggs of Liverpool. He was possessed of excellent abilities, which had been carefully cultivated by his parents from infancy. From the Charter-house he went to Cambridge, where he was unanimously elected to the Tancred medical studentship of Caius College, which situation he held for eight years, passing the whole of the time in colleges, hospitals, dispensaries, and dissecting-rooms, in Edinburgh, London, Dublin, Paris, and Munich. In each university he was distinguished by his unremitting diligence in the acquisition of science, and equally so for the amiable qualities which adorned his mind and made him a general favourite wherever he was known. During our tedious voyage, and especially since our entrance into the Niger, to me his society had been a source of constant grati-

fication and delight; and my obligations to him
for his judicious advice and kind attentions I can
never sufficiently express. The uncommon equanimity of his temper, the total absence of all
selfishness, and his desire to make the best of
everything, rendered him a most invaluable
companion at all times; while his prudence and
judgment were of essential service to me on
many occasions.

It was a bitter addition to my sorrow, that I
had been instrumental, in a great measure, in
bringing him out with us. When I determined to
accompany the expedition, I wrote to him, then
at Cambridge, offering him the appointment of
medical officer to the expedition. He had received my letter while at commons, retired from
table, and accepted the offer by return of post.
I knew his value at a sick-bed, as he had visited
and nursed me when I was labouring under a
malignant typhus fever in Edinburgh, in 1829;
I was acquainted with his value as a friend, as
I had known him from infancy,—as a man of
science, for I had witnessed the high ground
he had taken among men of his own standing;
and I fondly hoped that we should have returned
home with our friendship more firmly cemented

by our sufferings and adventures in each other's
company, the remembrance of which would
then serve to add an additional zest to the enjoyment of civilised society.   His remains were
deposited in a lovely spot on the eastern bank
of the river, by Hector and Sarsfield, the only
white men who were able to pay him this last
sad office.   It was a place to which he had often
resorted when well, and one which will ever remain sacred in my memory.

If my condition had been monotonous and tiresome hitherto, it was now doubly so; and, anxious to quit a scene which kept alive so many
distressing recollections, as soon as I recovered
my composure, I hurried on the preparations
for my voyage up the Shary, and in the latter
end of March, taking with me seven Kroomen,
two boys, and Thomas Sarsfield, I was lifted
into the boat.   A hurricane-house, which was
water-tight, had been built up abaft, sufficient
to hold two persons very comfortably; and a tar-
paulin cover had been prepared for the Kroomen
and the goods.   I left Harvey in charge of the
vessel, and Hector in charge of the goods, having great confidence in both, and departed on
my voyage to Fundah.

In a few hours I reached the Alburkah, and found that Mr. Lander was living on shore in a hut which the Kroomen had built. I was carried there, and was sorry to find Mr. Lander much reduced by dysentery. I had not seen him for six weeks, and we both had a great deal to say to, and to hear from, each other. Another quarrel, it appears, had taken place with the natives, who had attempted to attack Mr. Lander's hut, and in the skirmish that had ensued several had been wounded. Provisions were scarce, and ivory scarcer. This did not look very cheering, and I advised Mr. Lander to drop the Alburkah down to the Quorra, as the natives were not quarrelsome there, and we had a sufficient supply of provisions brought alongside; or else, that he should make another attempt to get up to Rabbah. I remained on shore two days, and certainly never envied a man so much the use of his legs as I did Lieut. Allen, who was in excellent health, and riding out morning and evening. Captain Hill was unwell, and remained on board his vessel: he appeared dropsical and low-spirited. I gave him some good advice and some clothes — the latter of which he accepted, but I doubt whether he profited much by the former.

Finding that Mr. Lander was unwilling to move in the state he was in, I determined on proceeding alone to visit Fundah, of which town we had heard so much, and so many favourable accounts.   Mr. Lander promised to follow me up as soon as a remission of his complaint would allow him, and gave orders to my mate, Harvey, who had accompanied me to the Alburkah, to get the cutter fitted with a hurricane-house for the purpose.   I engaged a man belonging to a neighbouring village to pilot me to Yimmahah, which was represented as the port of Fundah, for six thousand cowries, and started on my expedition in the early part of a delightful morning.

The main branch of the Shary is from three quarters of a mile to a mile wide at the confluence; but after a vain attempt to find a passage through the numerous sandbanks that choke its mouth in the dry season, we were obliged to re-enter the Niger, and ascend it for about two miles; when we came to a narrow creek, with three or four fathoms of water in it.   This creek intersects the rich alluvial soil thrown up by the united streams of the rivers in the form of a horse-shoe.   This delta is also intersected by numerous other creeks.   After passing the Al-

burkah, I was very particular in directing soundings to be taken, and we found two fathoms
where Captain Hill had reported that he could
walk across. Either the bed of the river had
sunk, or he must have been mistaken in his soundings. On re-entering the main stream, we found
an average depth of two fathoms, and the current very rapid ; — indeed so rapid was it, that
the Kroomen could not stem it with their oars,
and were obliged to " track " the boat, in nau-
tical phraseology, or tow her by a line.

After a tedious navigation of seven days, we
arrived at Yimmahah, having passed a considerable number of villages on both banks. At
one of these, called Fundykee, I was obliged to
pay the chief a small due, which he levies on all
canoes passing his town. Here I discharged the
man whom I had engaged as pilot; paying him
his cowries, and giving him a good sound flogging for imposing on me, as it appeared that he
knew nothing about the river.

In the evenings I generally came to, off a village, and after supper allowed the Kroomen to
go on shore with their drum, which seldom failed to set the heels of the whole population in
motion. We obtained plenty of provisions from

the villagers, who were always delighted to see
us, and to feel a white man,—for feel him they
must, before they will be satisfied that he is not
whitewashed for the occasion.

The course of the river to Yimmahah is E.
half N. by compass. It was now the height of
the dry season; but, as far as I can judge, there
are two fathoms' water in the channel, which is
dangerous only from its sudden windings and
rapid current at this season.

Yimmahah, where we arrived in the evening,
is beautifully situated on the top of nearly a perpendicular rock. At the time that we visited it,
a sandbank extended about a mile from its foot,
at the extremity of which were erected a great
number of temporary huts, the inhabitants of
which were occupied in catching and drying fish.

On my arrival, the chief of the village came
down immediately to welcome me, and informed
me that he had given up part of his own house
for my accommodation. Being tired of the confinement of the boat, I accepted his invitation;
and was carried on shore in a hammock suspended from a pole, which was borne by two of
my Kroomen. We got on well enough until we
reached the foot of the hill on which the town is

Engraved by W Finden A.R.A. from a Sketch by I. Sarsfield.

TOWN OF YIMMAHAH ON THE NORTH BANK OF THE SHARY.

situated, the only approach to which is by an almost perpendicular ladder of rocks.   By lashing the hammock to the pole, they managed to carry or rather to hand me up, but not without damage, as I was severely bruised about the shoulders and hips.   On arriving in the town I was much exhausted, and for the first time since my sickness, a period of nearly five months, took some brandy.   I had abstained from both wine and spirits, or other stimulants of any kind, from a mistaken notion that they were injurious : I am convinced that on that night a strong dose of brandy and opium saved my life.

In the morning I despatched a messenger to Fundah, distant about thirty miles, to inform the king of my arrival, and desiring him to send an escort and carriers for my goods.

Yimmahah is situated on a hill which naturally is almost impregnable.   It is separated from the main land by deep ravines, and presents a perpendicular face to the river, accessible only by the natural ladder of rocks to which I have alluded.   In addition to its natural defences, it is protected by a wall and ditch ; and on the eastern side a causeway is built or thrown up over the ravine, with a temporary bridge of wood across the ditch from the gate of the town to the causeway. The town contains about three thousand people; but the chief assured me, that before it was burnt by the present King of Fundah, it contained ten thousand inhabitants. A romantic kind of story, which I heard here of this King of Fundah, did not give me a very favourable impression of his character. The account said, that he is the youngest of five brothers, two of whom he poisoned; and another, in order to avoid his cruelty, had committed suicide: the remaining one fled to this town, and was beheaded by his brother, after a war that continued for some years, and by which the country was desolated, all trade destroyed, and the caravans from Kano and Bornou to this part of the country interrupted. I do not vouch for the truth of this story, although it was confirmed by several people afterwards: I only " tell the tale as it was told to me." I also heard various accounts of the King's rapacity and cruelty towards traders; but, as I had gone so far, I was determined to proceed, feeling confident that the accounts were much exaggerated, and being anxious to visit a city that had only been heard of, like the mysterious Timbuctoo, through the obscure and contradictory accounts of the natives.

During my stay at Yimmahah, I was treated with the greatest deference and respect by the inhabitants, and at last got so intimate, that the children would even shake hands with me. An old Mallam, who was the owner of some extensive dyeing works, and reputed the richest man in the place, was my constant companion, and from him I learnt much of the state of trade in the interior. He confirmed my opinion, that formerly there was a direct communication between Kano and other towns in the interior and this part by Cafilas, and that the rapacity and extortion of the King of Fundah had put a stop to it. I must do this old gentleman the justice to state, that he strongly dissuaded me from going to Fundah, saying very truly, " If the king has ivory, he can send it to you here, and not trouble you to go to him."

In the course of my stay at Yimmahah, my attention was much struck by the rapidity with which they thatch their houses. The process is simply as follows :—A pole is placed upright on the ground to form the apex of the roof. A circle is traced on the ground by a string from this

pole ; and the framework of the roof, extending
from the top to all parts of the circle, is tied
together with grass ropes and interwoven with
twigs. It is then raised and placed upon the
walls of the house for which it is intended; and
the thatch, which is composed of grass carefully
dried, is laid on in a manner equally as neat as
that of the best thatched English cottage. I
have frequently seen a roof of this kind, and
considerable in point of size, begun and finished, thatch and all, in two days. The tops of
the principal huts are generally ornamented with
a ventilator of a fantastic shape, surmounted by
a straw figure of a cock.

The hut in which I resided was the entrance
to the chief's apartments, which consisted of
seven separate huts connected by a wall, one of
which was appropriated for our goods, and the
rest were occupied by his family. While remaining here I had my head shaved, or rather clipped,
and my beard trimmed, by a country barber, who,
like his brethren in other parts of the world, had
a vast deal to say for himself. I learned from
him, that I was in a fair way of being the lion of
the day in Fundah,—that the whole town was
on the tiptoe of expectation, and that he himself

had come over on purpose to see me, in order to gratify the curiosity of his customers. He was an expert operator, and clipped my head with a pair of scissors of native manufacture nearly as close as if it had been shaved, besides shaving the upper part of my face with a razor of the country. He used country soap, which produced a strong and agreeable lather; and I was exceedingly amused at the reverence which he paid to my beard during the whole operation, frequently indulging in expressions of admiration of it, and assuring me with the utmost earnestness that there was not such another in the country.

CHAPTER VIII.

Messengers from the King of Fundah.—Embark in a Canoe for that City.—Difficulties of the Passage.—Arrive at Potingah, and pass on to Fundah.—Native Curiosity.—Visit from the King.—The Visit returned.—Inconvenient Lodgings.—The Court of Fundah and its Inmates.—Sarsfield sent to Yimmahah.—The Barber Physician and his Wife. —An Amateur Beau.

ON the 4th of April, to my great relief, the messengers of the King of Fundah arrived. They were eighteen in number, and came galloping into the town on small high-spirited horses. Instead of coming direct to me, they went to a house at some distance, and sent a messenger to say that they would wait on me. Shortly after, they made their appearance, galloping up in rotation to the doorway and there dismounting; they each bent down and put my foot on their heads, and threw dust over themselves. Having finished this salute, they became seated, and their leader opened the palaver by saying that the king his

master had sent him, his chief warrior to escort
me to his capital, where I should find everything
I wanted, &c.; and that, lest I should starve
by the way, he had sent me a dash of some
goats and yams. My reply was in presents; but
I expressed at the same time my annoyance at
the delay which had been shown and my detention thereby, and that as I was unable to sit on
horseback, I should require bearers for my hammock.

This, however, I found much easier to ask
than to obtain; and although I had concluded
our palaver with the request, and expended three
more days in further palavering, it was all in
vain — no bearers appeared. However, I discovered that in a small canoe I could get within
nine miles of Fundah, and from thence be carried there by my Kroomen. I therefore consented to compromise the matter with them,
and determined that Sarsfield should go overland with them in charge of the goods I intended to take with me. I sent my Kroomen also
with him, armed with muskets and cutlasses, in
case of treachery, each having twenty rounds of
ball cartridge. I left two Kroomen in charge of
the launch and part of the goods; and with the

remainder I embarked in a canoe, accompanied by one of the black boys as interpreter.

Although I had found it difficult enough to get into Yimmahah, I had no trouble in getting out of it, as the old Mallam mounted me on the shoulders of one of the natives, and telling me to hold on by his ears, led the way down the precipitous face of the rock, my carrier following him and leaning on his shoulders. I shut my eyes and held fast, and was most happy when we had reached the bottom safely, as a single false step would assuredly have been fatal to both of us. My bearer trotted across the sandbank with my long legs dangling round his neck, amidst the laughter of his companions, great numbers of whom had assembled to witness my departure. The canoe which I embarked in was about twenty-five feet long, pulled by four men, who, with one of the Fundah messengers, completed my crew.

We started at midnight, and had proceeded only a short distance when, to my great surprise, the canoemen pulled in to the shore, landed, and deliberately left me in the canoe with my boy while they repaired to some huts for the purpose of sleeping. It was in vain that I remonstrated with them on this proceeding, asserting that they had agreed for additional pay to proceed during the night, that I might avoid the dreadful heat of the day;—I was completely in their power; and as there was no alternative but that of submitting, I desired the boy to make a large fire on the bank, and, lighting my pipe, contrived to smoke myself into something like equanimity. In the morning the canoemen returned; and as grumbling was of no use, I endeavoured to coax them into a little extra exertion by purchasing provisions for them.

As we proceeded up the river, I was much struck by the size of it. We passed immense sandbanks, which as they are covered when the river is swollen, it must then afford a magnificent prospect. In the afternoon we arrived at a small creek, which we had great difficulty in entering, and again stopped for the night. My crew this time had the civility to pass their night's rest close to me, taking the precaution to make two immense fires to keep off the wild beasts, which, they said, were numerous in that neighbourhood. One of the crew also kept watch, and was relieved in the night by his companions.

In the morning we again proceeded. The

small river into which we had entered runs in
a north-north-east direction, and is exceedingly
shallow. Twice we were obliged to land the goods,
and drag the canoe over a kind of rapid with a
sandy bottom. The heat was excessive, and I suffered much : my face was burnt terribly, and my
hands and feet were blistered and swelled. The
canoe was without cover, and my exposure to
the full power of the sun literally distracted
me with pain. In vain I intreated the canoemen to pull in towards the side of the river
that we might be sheltered by the trees : the
boy that I had as an interpreter became frightened, and would not speak; and the men, I believe, thought me deranged. A few hours more of
such suffering would have made them right, for
assuredly it would have brought on a brain fever.
The agony which I endured on that day, I can
only compare to the sufferings of a person roasting before a slow fire. My only relief was by
wrapping myself up in a thick blanket; and this
was a miserable alternative, the feeling of suffo-
cation being about as bad as that of burning.

In the afternoon we reached Potingah, where
I had the satisfaction of finding Sarsfield, who
had arrived the day before. He had been much

alarmed at my not having appeared sooner, and was overjoyed to meet me again.   He had received the greatest civility from his escort, but, like most sailors, had experienced quite enough of horse exercise.   After taking some coffee, I went into my hammock, and, accompanied by a cavalcade of about twenty horsemen, set out for Fundah, distant about nine miles.   My Kroomen were obliged to carry me, as the natives objected to carrying a white man, alleging that they were not horses.   I offered three times as much as I gave them for carrying the goods, but to no purpose; they still repeated that they were men, and not horses.

The road from Potingah to Fundah is nearly due north, and, with the exception of three ravines with a little water in the bottom of each, is an excellent bridle-road.   These ravines are impassable in the rainy season, and a detour is then made to the northward, the road running along the declivity of a range of hills, which, beginning at the confluence of the Shary and Niger, runs parallel to the river at the distance of seven or eight miles.   The country was cleared and appeared well cultivated.   On emerging from a ravine, we entered the plain on which

the town stands, and having arrived within a
short distance of its walls, we halted until the
king was informed of our approach.

It was midnight, and a splendid moon threw
every object into bold relief. Even at this late
hour an immense crowd had assembled outside
the walls, quite in corroboration of the barber's
evidence as to the state of public curiosity. After waiting some little time, during which we
were visited by several Mallams, we entered the
town, and were conducted to a miserable hovel
in the centre of it, accompanied by a mob of both
sexes desirous of seeing and touching a white
man. I cannot say much for the civility of our
first reception. It was nearly daylight before we
got rid of our tormentors, and were allowed that
repose which the fatigues and excitement of the
day had rendered so necessary. Among other
annoyances, they thrust a disgusting Albino close
to me, and asked if he was my brother!

In the morning we were removed to a hut in
some respects superior to our first abode, but
unfortunately situated, inasmuch as it was open
to a wide street, which was filled from daylight
with a crowd whose curiosity it was impossible
to satisfy, and from whose tongues proceeded a

Babel of noise and clamour which it is quite laughable to think of now, although it was sufficiently annoying at the time.

About ten in the morning, the king's chief eunuch came to see our packages of goods examined, in order that we might be satisfied that our conductors had taken nothing. On finding everything right, I made a present to each of the escort and bearers far beyond their expectations. In the afternoon I was visited by the king, who was attended by a great number of eunuchs and a cavalcade of about a dozen horsemen. He was splendidly dressed in silk and velvet robes, and appeared to be a man of immense size. His countenance is by no means prepossessing, particularly his eyes, which are of a dirty red colour, having a sinister and foreboding expression. I presented him with a brass-mounted sword, an umbrella five feet in diameter highly ornamented, a brace of pistols, and several other things, and then informed him through my interpreter that I had come from a great distance to look at him in the face, and to hold a good palaver with him; that his messengers had informed me it was his desire to see the face of a white man, and trusting to his good faith, I

had come, though ill and unable to walk; that I
was anxious to give him our goods for ivory, and
had brought with me a great quantity for that
purpose.

Having finished my speech, he rose, and said in
the Houssa language, that he was glad to see the
face of a white man—it was what he had long
wished for; that he had abundance of ivory, and
that all that he had was mine: to which sentiments twelve grey-headed negroes, who appeared
to form his privy council, bowed assent. I then
complained of the reception which I had met
with, the miserable hut in which I was lodged,
and requested that I might be provided with
another in a more private situation immediately.
To all this he readily assented, and moved off
amidst a horrible din of trumpets, drums, and a
kind of fife-music. — But I had yet to learn
a lesson: it seems that everything which the
nether end of this royal personage touches is
sacred, and has the *broad arrow* affixed upon it.
As soon as the king rose from his seat, which most
unfortunately happened to have been the tin-box
that contained my clothes, covered with a Turkey
rug, his attendants deliberately shouldered my
unfortunate box, and strode off after their mas-

ter. This was rather too much, and before they had got out of the yard, they were happily stopped by two of my Kroomen, and after a long altercation gave up the box, but detained the rug.

In the evening I had a visit from a man whose face I thought was not new to me, and a lady who assured me she was the king's mother, and to whom it was intimated that I should give a present. A looking-glass and a cake of Windsor soap satisfied her, but not her companion, who became abusive, and was at length bundled off by my Kroomen. A most horrible noise prevented me from getting any sleep until nearly daylight : my boy Friday told me in the morning that it had been occasioned by a dance at the king's house, which was close by ; adding, that as such was a favourite amusement, it might be expected frequently.

On the following morning, I was carried to the king's house to return his visit, but was only allowed to enter the outer court-yard, which is about forty feet wide, with a verandah on the side next to the house. Under this verandah I was placed, and in a short time the very man who had been turned out of my hut by my Kroomen the night before came and sat down by my side. After some conversation, I asked for the king; on which he said that he was the king! This was too much for me to believe, until he went through a gateway and returned in a few minutes with his stomacher and his splendid robes on. After laughing heartily at my astonishment, he asked for the carpet on which I was seated, and which I refused him, having no other. After some angry words on both sides, he went off in a pet, and I returned to my hut in any but a pleasant state of mind. On inquiry of the owner of my hut, he informed me, and I afterwards found it to be the case, that on all great occasions it is customary for the king and his attendants to puff themselves out to a ridiculous size with cotton wadding; and this fully explained the mistake I was under with regard to the king's identity. On his first visit he appeared to be an immense-sized personage, and could not even rise from his seat without assistance. When he visited me *incog.* he was a raw-boned, active-looking man.

Sarsfield, who had been hunting over the town for a desirable house, informed me that he had found an excellent one, belonging to a Houssa woman, and I determined to take possession of it, as there was no appearance of the king's supplying us with another. In the morning I was carried up to my new lodgings, which I found clean and roomy in comparison with those I had left. The proprietor of them was a widow who was reputed to be rich, and she welcomed me to them with much kindness.

I had just time to congratulate myself on the change, when Sarsfield came hurrying in with terror and dismay in his countenance, saying that the king had seized the goods that were left at the other hut—that he had taken them all into his own house, and was threatening vengeance on us all for daring to change our residence without his permission. This was confirmed by my hospitable landlady being seized by some eunuchs, and forthwith carried before the king to be punished for receiving us. As I considered all this a mere pretext for seizing our goods, I determined on remonstrating with him personally on this breach of hospitality and good faith. I was accordingly carried down to his court, and took my seat under the verandah, near my landlady, who was crying bitterly and tearing her hair, having been put in irons preparatory to being flogged.

After waiting a quarter of an hour, in rushed his sable majesty, and coming up to me commenced a volley of abuse, which it was perhaps as well that I did not understand, as I was already sufficiently exasperated against him. In the course of his hurried harangue, he repeatedly drew his hand across his own throat, using at the same time various significant gestures, with the view, I concluded, of intimidating me. But I had been too long among negroes, and too well initiated in their character, to feel the least alarmed at his proceedings; and just as he concluded, Sarsfield brought in my Kroomen all armed. They had no sooner made their appearance than a change was produced in the tone and manner of the king; he became at once moderate in his proceedings, and the palaver happily ended in a more peaceable way than its commencement seemed to promise. It was agreed that I should give up my new lodgings, and remain within the outer walls of his house in two huts and a large court-yard, beyond which I was not to go without his permission. In return for this compliance on my part, he consented, though with a very ill grace, to release my unfortunate

Engraved by W Finden A.R.A from a Sketch by I Sackfield.

ACCOMODATION AT THE COURT OF FUNDAH.

landlady, who, I think, will require not a little persuasion before she again lets her house to a white man.

As I was now established in the court of the King of Fundah's palace, I may describe it.—It consists of nothing more than an immense assemblage of circular huts, surrounded by a mud-wall about fifteen feet high and three feet thick. Through this wall there are three principal gateways, besides several doorways which open into an outer court: beyond this court strangers are not permitted to enter, but, from the account of my boy Friday, who was in the interior several times, it is merely an open space surrounded by small huts, in each of which reside one or more of the king's wives. The whole is in the form of a horse-shoe, and covers a space of about nine or ten acres. The part allotted to us consisted of two huts of about twenty feet diameter, with a verandah projecting about five feet from the walls, and a court-yard about thirty feet wide. A door was broken in the outer wall for our use, and the doors that communicated from the huts to the inner space were carefully stopped up with brick. Our court-yards were separated

from the king's by a wall, but a doorway allowed
free communication.

The huts and verandahs allotted to us were
in bad repair and swarming with vermin of all
sorts: but by hanging my hammocks between the
posts that supported the verandahs, I kept tolerably clear of those that crawl upon the face of the
earth : to the winged annoyances I was pretty
well inured.

After being in my new quarters a few days, I
began to suspect that my Fundah trip would
turn out an unprofitable speculation, as two or
three natives had been severely flogged for attempting to sell me some ivory, and the king,
who visited me every day, evidently had none,—
or if he had, thought it better to keep it and my
goods also. I had, in short, been completely
decoyed, and had only myself to blame, as I had
put myself entirely in the king's power, and, from
my excessive debility and emaciated condition,
was evidently becoming an object of contempt
to men who pride themselves above all things
on their personal appearance, and also consider
rotundity of figure as a sure sign of rank and
condition. In addition to my skeleton-like appearance, I was suffering under a complaint

called craw-craw, with which I had been infected
in the small canoe when on the way from Yimmahah to Potingah.   Compared with this the
disorder called the " Scotch fiddle " must appear
a pleasing titillation, as, when a fit came on, it
threw me almost into convulsions : this induced
a nervous irritability, which in my weak state
was at once pitiable and laughable, and tended, I
have no doubt, to lower me in the opinion of the
natives.

Sarsfield, who enjoyed excellent health, and
consequently much better temper, was comparatively well treated, but, having a smooth chin,
was considered as a boy.   Our time hung heavily on our hands, and a description of one day's
proceedings would nearly serve for all :—At daybreak we enjoyed about two hours' delightful
repose, between the cessation of the plague of
mosquitoes and the commencement of that of
flies.   At eight o'clock the king sent us our
allowance of yams for the day, and the process
of cooking them for our breakfast whiled away
an hour or two.   From that time until sunset
we had a succession of visitors, whom the king
admitted through the gateway, and, I verily
believe, made a regular charge for it.   To see

us eat was the summit of curiosity with the
natives; and if they could manage to work me
into a passion,—which, I am ashamed to say was
too easily done,—their delight was beyond measure. After being subject to this annoyance for
some time, I threatened to shoot any one who
came into my court without my permission; and
the king seeing I was in earnest, stopped the
street that led down to my quarters, and during
the remainder of my stay no one could enter
without my being made aware of it.

After sunset the most diabolical noises commenced in all quarters of the town, but particularly in the court next to mine, where the
king's musicians, as they called themselves, performed every night. The band was made up of
native drums, fifes, triangles, and trumpets, and
effectually prevented me from obtaining any rest
until two or three in the morning: and if to this
be added the annoyance from myriads of mosquitoes, whose everlasting hum was to me ten
times more painful than the severest bite; the
constant screeching of those African scavengers,
the Turkey buzzards; the legion of ants, which,
justifying Solomon's praise of them, are constantly on the watch for anything living or dead;

the squadrons of enormous rats, that immediately after sunset commence their evolutions beneath one; and the croaking of thousands of bull-frogs from a neighbouring wood, some idea may be formed of the pleasures *attendant* on a visit to an African court.

I saw a great deal of the king; indeed I had far more of his company than I desired. He generally came to me in the afternoon, attended by several eunuchs and boys, who carried the keys of his various huts. His occupation, when with me, was to ridicule my weakness, and mimic my attempts to walk; for I had so far recovered as to walk a few steps with the assistance of two sticks. I cannot describe the disgust I had of this man's presence; and if he had persevered in his behaviour, I should certainly have shot him in some moment of irritation, as he stood before me. To my repeated applications for horses to remove myself and my goods, he only replied that his gods would not allow him to part with me. It was with difficulty that I prevailed on him to permit Sarsfield and the Kroomen to return, under pretence of bringing more goods from the vessels. I kept a Krooman named " Pea Soup ;" a mulatto, named Smith, who was ill of dysentery, and my two boys, as my attendants, and gave Sarsfield directions to proceed down the river, and return as soon as possible with some articles I was much in want of. My small supply of coffee had been long exhausted, and I was in want of some medicines for Smith and myself. He was also directed to bring some rockets and blue lights, and to arm the Kroomen with pistols and cutlasses as well as muskets. I wrote by him to Mr. Lander, stating my situation, and begging him to bring the Alburkah a short distance up the Shary, in order to receive me if I succeeded in getting away, but on no account to allow the white men to return with Sarsfield. I had been rather surprised at Mr. Lander not coming up after me, as he had promised, but concluded that he still continued indisposed.

Sarsfield set out in good spirits at day-break, and, although the king had promised him a horse, he refused to let him have one, in consequence of which he was obliged to walk to Yimmahah, and I was left alone to " chew the cud of sweet and bitter fancy" until his return. Smith, the mulatto, was very ill with dysentery ; and the day after Sarsfield's departure I was seized with the same complaint, and for several days was exceed-

ingly ill.   A native disciple of Esculapius cured
me with raw rice ground very fine, and mixed
with water.   After drinking great quantities of
this mixture the complaint gradually ceased, and
I astonished my physician by giving him a fee,
which, though trifling in itself, was to him an
invaluable treasure, — it was merely a case of
razors,—for the physician was no other than my
friend the barber.

Smith would not take his medicine, and was
very ill.   I heard of Sarsfield's arrival at Yimmahah, and his departure down the river, from
my doctor, who was the very " Times " itself of
Fundah ; and his wife was no less communicative.
The rice which I took was mixed up and brought
to me by this lady, who, to prove its harmless-
ness, was in the habit of stirring it round
with her fingers, and drawing them through
her mouth before presenting it to me to drink.
However, she was a good creature, and I may
here add my testimony to that of all African
travellers to the uniform kindness of the women.
In my case the sentiment of pure compassion
was unalloyed by any personal consideration, as
I was anything but an agreeable object, and had
a most irritable temper ; yet, at the hazard of

being flogged if discovered, they supplied me
with eggs, fowls, and sometimes butter-milk,
and never did I receive from them anything but
kindness and pity.

My tormentor visited me daily during my illness, which lasted about eight days, and I could
not help thinking that he was watching like a
vulture for the moment that my breath left my
body, to be able to pounce upon my goods. I
was not at all afraid of his taking them by force,
as he knew that I had two barrels of gunpowder;
and I had told him, that if he attempted such a
thing I would blow his whole town about his
ears; nor was I apprehensive of his personal violence, as I felt convinced that he dare not attempt it, from his superstitious dread of the
power of a white man; but I must confess that
the thought of being poisoned haunted me incessantly: it is a horrid, disagreeable, sickening
kind of feeling, the idea of being poisoned; but
unfortunately, it is very commonly practised by
the chiefs, and suits the cowardly nature of
tyrants. I used great precaution, taking care
always to cook my own victuals, as, after Sarsfield had left me, I had no one in whom I could
place any confidence. My boys were made so

much of by the king, that they both became perfectly independent of me; and knowing my inability to move, and keeping carefully out of my reach, would rob me of cowries before my face, to supply their extravagance. Friday, who was a good-looking lad, was quite a lady-killer, and would take himself off every morning to the barber's shop, to get his black woolly head dressed in the most recent fashion (an exquisite of Fundah must have his hair dressed every day). Mr. Friday's next occupation was smoking his pipe and relating stories, wonderful no doubt, about Sierra Leone and the Havannah, to the admiring crowd of natives, who would collect about him. His amourettes repeatedly got him into scrapes; but " the more danger the more glory," seemed to be his motto, and he was altogether incorrigible, although heavy damages were frequently levied on his shoulders, which in England would have fallen on his purse.

CHAPTER IX.

THE rainy season was now fast approaching, and tornadoes occurred almost every night. The thatch of the verandah under which I lay admitted the rain in all directions; and having no water-proof covering to my hammock, I was frequently soaked before morning. I did what I would recommend others to do in a similar situation, that is, to smoke continually, and take large doses of opium; and at sunrise spread everything out to dry. After one of these tornadoes I had a narrow escape, which I shall not easily forget. It was my custom, at sunrise, to be lifted out of my hammock, and laid on a

mat while my blankets were dried. One morning, on calling my boys, they discovered a small
black snake coiled up beneath my hammock,
and, telling me of it, and that it was exceedingly venomous, off they scampered to fetch some of
the natives. I looked over the side of my hammock, and about three feet below me lay the creature, about a yard long, with his head in the centre of the coil, his eyes peering about, evidently
bent on mischief. For two hours we remained in this enviable position, while about twenty
of the natives were grouped in the court-yard,
chattering, and pointing to the snake, and I not
in the best of humours alternately entreating
and threatening them for not removing it.
My good genius, however, appeared at last in
the shape of an old woman, who killed it, by
piercing it to the ground with a forked stick.
The natives assured me it was most venomous,
and, from their dread of it, I fully believed it
was the case. It was the first and only venomous creature I met with in the country.

I had begun to be seriously alarmed about
Sarsfield, when to my great relief, he arrived on
the fourteenth day after his departure. To my
utter astonishment, I learnt from him that Mr.

Lander had gone down the river in a canoe!
with Captain Hill of the Alburkah, without informing any one of his intention : that Lieutenant Allan, had had an attack of brain fever,
and in his delirium had stabbed the cook of the
Alburkah, who was not expected to recover :
that the Alburkah had lost one seaman by
dysentery, and that the Quorra's crew were
better.   All this was confirmed by a letter from
Mr. Hector, whom I had left in charge of the
goods, and who, when Sarsfield arrived, was on
the point of coming up the Shary in search of
me, as the natives had informed him that I was
detained at Fundah. This information annoyed
me very much; not that I cared whether Mr.
Lander went to sea or not, but I had lost an
opportunity of communicating with my friends
and family at home, and preventing them from
taking any steps or embarking more deeply in
the speculation, which erroneous information
conveyed directly or indirectly might lead them
to do.

If I was glad to see Sarsfield, he was overjoyed
to find me still alive, and though much fatigued
with his journey, having walked from Yimmahah
a distance of thirty miles, we sat up until nearly

daybreak, and laid a plan for frightening the king and his people, all the credit of which is due to him both for its invention and success. He had brought the rockets with him, and had let off one at Yimmahah, and described the terror and alarm of the inhabitants produced by it as excessive. They looked on him as a deity, and supplied him in consequence with provisions and carriers to Fundah. He proposed to try the effect of them here, letting off three or four at a time, and burning blue lights after them.

The next morning we had a visit from the king, who wished to see what Sarsfield had brought. I gave him to understand that my people would not send anything until I went to them myself, and that in the evening I intended to make a grand fetish to my god, to know whether I should go, or stay at Fundah. The king said that was good, and that he would attend with all his priests and summon the inhabitants to witness it. The fetish was to be made under a large tamarind-tree, at the upper end of the street in which the court was wherein we resided. We made as much of this affair as we could, and pretended to go through sundry preparations, in order to impress on the minds of these people an idea of its importance.

In the evening I was carried out about seven o'clock, and seated in the street opposite the tree, the king and his chief men close by, surrounding Sarsfield and the Kroomen, who were holding the rockets and blue lights that we had brought out for the occasion. As a commencement of the proceeding, I took a piece of paper and with great solemnity fastened it to one of the rockets and gave it to Sarsfield; we agreed that the signal for firing the rockets and blue lights, should be the discharge of my pistol. An immense crowd of natives was assembled to witness the ceremony of the white man's fetish; the wide street was filled, and the roofs of the houses and tops of the walls were crowded with spectators, all full of wonder and speculation as to what they were to see.

Everything being ready, I fired my pistol, and up flew four beautiful two pound rockets, the discharge of which was immediately followed by the blaze of six blue lights, throwing a ghastly glare over the whole scene. The effect was perfectly electric; the natives had no idea of what was coming, and fled in all directions. The king,

filled with terror, threw himself on the ground
before me, and placing one of my feet on his
head, entreated me to preserve him from harm,
and to inform him what was the decision of the
Fates. It was now my turn to make use of the
power which I was supposed to possess, and I
replied that I should tell him presently, but that
I must now return to my house with all my men.
The farce had been successful so far, and the
artifice was only to be carried out to a successful
issue.

After keeping the king in suspense about an
hour, I sent word to him that I was ready to
receive him, and that he himself was to come
and see the result of the fetish. He came immediately, and as soon as he was seated, I told him
that I had sent for him to see whether I was to
go or stay, and that my god would punish them
in a manner of which they had no idea, if they
presumed to break his commandments. I then
took from my pocket a little compass, and explained to him, that if the needle in it pointed
towards me, as soon as it was placed on the
ground, that I was to go from Fundah, and that
if towards him that I was to stay, to all of which
he attended with much reverence. I of course

took care to know my position, and placing the
compass on the ground, the needle very properly
turned towards me. This was sufficient, the
thing was done, and the king was convinced
that he should allow me to depart freely. I
wished him to handle it, but he shrunk from it
with terror, imagining that it was a living creature, and was glad to get out of my presence,
after having promised to give me horses, or what
ever I might want in the morning. We had
the greatest difficulty to preserve our gravity
throughout this farce, and when left to ourselves
we enjoyed a hearty laugh at the success of our
plan and the credulity of the king.

The next day was passed in joyful preparations
for our departure, and with the exception of the
heavy goods, and the two barrels of gunpowder,
every thing was packed up for the carriers. At
daylight the next morning, I was lifted on a
horse, being anxious to get to Potingeh before
the sun was high; and, leaving Sarsfield to bring
the goods, I turned my back on a city where
I had suffered both bodily and mentally more
than I could describe, and from whence I took
nothing but experience dearly purchased by an
expenditure of time that would have allowed me,

if I had ascended the Niger, to have reached
Boussa. I was accompanied by a host of the
inhabitants, whose curiosity, although I had been
seven weeks among them, had not undergone the
least abatement.

As the town of Fundah is generally allowed
by the traders to be the largest in that part of
the country in which it is situated, and as it has
always been considered by geographers and former travellers, as a place of importance, both
from its trade and position, a short account of
it by its first European visitor may not be misplaced, and will naturally be expected.

Fundah is situated at the western extremity
of a magnificent plain, on the northern bank
of the river Shary, from a branch of which it is
distant about nine miles. It is bounded on the
north by a range of low hills, which, commencing
at the confluence of the Shary and Niger, run
in a direction parallel to the former. To the
eastward of the town a beautiful country extends
as far as the eye can reach. The town itself is
built in the form of a half moon, walled in,
on three sides surrounded by a ditch, and on the
other it is defended by a low wall and fence
which has again a natural defence in a deep

ravine. The height of the wall round the town is about twelve feet, and its thickness about six, and the depth of the ditch is about ten feet. Sarsfield, at my request, walked partly round the walls; and his journey from the eastern to the western gate occupied about four hours. Allowing that he walked two miles per hour, that distance would give about two miles and three quarters as the greatest width of the town, which, I think, is rather under than over the mark. The only appearance of any art in the construction of the walls, was a flanking wall run along the edge of the ravine that bounds the town on the south side, evidently for the purpose of annoying any assailants. At the western point, this part of the wall had battlements, but the remainder had none. An open space is left inside the walls all round the town, no houses being built for a considerable distance from them. The king's house appeared to be the citadel, as it was surrounded by a wall which had loopholes in many places. The streets of Fundah are for the most part narrow and dirty, with the exception of that which surrounds the king's house; and from its principal gate, extends to the eastern gate of the town.

It is about two hundred feet wide, and about a mile long, and the market is held in it every Friday. The houses are all circular with conical roofs built of clay, with the exception of the chief Mallam's, which had a gable end to it. The verandahs in the front give them a cool and pleasant appearance.

It is difficult to estimate the population of Fundah. I may venture to say, that it contains from five to six thousand huts, and that the average of each hut cannot be less than six persons, two thirds of which are children. On the night that we made our fetish, in my opinion, there were twelve or fifteen thousand people present, and I have frequently been carried through the main street when the crowd was so dense as to require my Kroomen to force their way; but this was on market days, when great numbers of people from the country were assembled. Nothing is more difficult than to state an accurate approximation to the population of a negro town. Most of the respectable inhabitants are Mahomedans, but nine tenths of the whole population are Pagans. The proportion of slave population to the free, as nearly as I could judge, is as five to six. The slaves generally are well treated, but badly and scantily fed. They would scramble in a disgusting manner for any morsels which we might throw them.— When they abscond, if they are recovered they are severely punished by flogging. I saw several with the marks of the cowskin upon their flesh.

The Government of Fundah consists "in might, not in right." The will of the chief appears law, but he has a council composed of very old men, with whom he pretends to consult. The present king is universally disliked for his cruelty and oppression; he is notorious for his bad faith, and by way of consolation in my experience of him, I was told, while living with him, that he had frequently enticed traders into his clutches, and poisoned them for the sake of obtaining their goods. The people freely expressed their abhorrence of him, and, from the feeling which I saw evinced towards him, I should think that his reign would not be long endured. He pretends to exist without eating and drinking, and I thought at one time that he wished me to follow his example, as I was very often unable to procure food of any description. Among his peculiarities, he has a violent antipathy to tobacco, and if he catches any of his subjects smoking, he breaks their pipes on their heads; if he sneezes, every one present must say, Salam Aleikum! if he laughs, his attendants must cover his face with their fans, to prevent his emotions being witnessed by the vulgar eye. At daybreak a slave arouses him from his slumbers with a loud voice, and he seldom retires until two o'clock in the morning. He is active and energetic, and exceedingly passionate, frequently knocking down his slaves with a single blow.

During my stay there was only one execution, if it can be called by that name, for it partook more of the nature of a sacrifice. An elderly chief died, leaving fifteen wives and no children. On the night the deceased was to be buried, the king went to the women's apartments and selected one who was to be hung, in order to accompany her husband to the next world, sending the rest to his own house. Sarsfield went to see the execution, which he said was performed in a very solemn manner: the people appearing to view it altogether in the light of a sacrifice. It is certainly preferable to the horrid custom which prevails at Old Calebar, Bonny, and other places on the coast, namely that of burying widows alive.

With respect to any force that might exist in Fundah, if any of consequence does exist, which I much doubt, it must be very insignificant. On two occasions, I was carried outside the walls into the lovely plain, at the extremity of which the town is situated, in order to witness a display of horsemanship, dancing, and racing, and where naturally the youth and chivalry of the town were assembled. But I do not think that they mustered altogether fifty horses, not ten of which were worth looking at twice.

A description of these amusements may not be uninteresting. I had been about a week in his house, when the king sent me a message, inviting me to accompany him to witness a fête which he was celebrating outside the walls. It was a beautiful morning; the heat was tempered by a pleasant breeze, as I was carried out by my Kroomen to the eastern gate of the king's house. After waiting a few minutes, listening to the most horrid noises that the yeomen of the king's band could perpetrate, out sallied from the interior square about twenty eunuchs as the advance guard; the king followed on horseback, surrounded by his musicians and body-guard of about fifty men, armed with bows and arrows, and six of them with muskets, (three of which were minus their locks,) accompanied also by his chief men and a number of Mahommedan mallams. The king graciously shook hands with me, and asked me with a smile of satisfaction, if ever I had seen in my own country such a sight as this; I answered him with great sincerity, that I never had; for certainly the wildest conceit of Ducrow never approximated in the slightest degree to the figure and appearance of his sable majesty and his companions. Their clothes were all stuffed so as to make them appear of an enormous size, and their horses being small seemed so completely overladen, that if I had not been aware of the want of solidity in their burthens, I should have wondered at their lively prancing under their apparently enormous riders.

I took the lead in my hammock, and after proceeding about a mile and a half from the palace, we halted under a magnificent wild cotton-tree, in the shade of which the king and his courtiers seated themselves. The whole scene was novel and inspiring; the walls of the town were crowded with spectators, and the road leading to the town was lined with an immense multitude of persons. The sport commenced

with horse-races: three or four horses started
from the gate, and galloping about half a mile
past us, they were pulled up with a jerk which
threw them on their haunches, when suddenly
wheeling round they came back, and their riders
made obeisance to the king. The racing was
succeeded by sham fights, the parties alternately
retreating and advancing; and the whole was
closed by a general dance, when the best per-
former was he who kicked his heels the highest.
During the whole time in which these things
were going forward the drums were beat, the
trumpets brayed, and the fifes squeaked, producing altogether a discordance of sounds utterly indescribable. My Kroomen afterwards
trotted out, and, with an air of conscious supe-
riority, exhibited their own national dances to
the accompaniment of a drum, before the won-
dering natives.

After the dancing, I was politely informed that
I had better withdraw, as the king was going to
offer up his devotions privately, in a small fetish
hut near us. I was accordingly carried back with
my curiosity unsatisfied respecting the ceremonies he was about to perform, and I never could
ascertain whether he was a Mahommedan or a

heathen. It is, however, most probable that he was alternately the one and the other, as he kept mallams as well as Pagan priests in his service.

The plain on which the races were held is covered with a rich verdure, studded with clumps of trees as far as the eye can reach. I thought at one time that I saw the Shary in the south-east quarter, but afterwards considered that it must have been a mirage. This plain affords a magnificent view from the walls of the town, and as my bearers halted at the gate, when I was going away, I could not help looking back, and sighed on reflecting that there was little chance I should ever be well enough to gallop over its free and wide expanse.

I imagine that the whole disposable force of Fundah was collected on these occasions, and I do not consider that there were more than three hundred footmen armed with bows and arrows, in addition to the thirty horsemen before mentioned. The park of artillery consisted of three serviceable and three unserviceable muskets; and as one individual alone was to be found bold enough to fire off one of these, which he did from the hip, it cannot be looked on as very formidable.

While residing at Fundah, I was repeatedly

entreated by the king to allow my Kroomen to
accompany his men on marauding expeditions.
My refusal irritated him exceedingly, and, I have
little doubt, was one reason for his conduct to
me being so harsh. He tried the men themselves; but though the Kroomen, like the Irish,
will fight for the fun of the thing, they knew
their duty too well to disobey me. They had
frequent wrestling and boxing matches with
the inhabitants, and were generally victorious,
although they never got into serious scrapes.
They were all tall, handsome men, and having
the advantage of the polish acquired by travel,
and the discipline of our cruisers, they surpassed
the Fundah beaux in the good graces of the fair
sex, who, frequently regardless of their reputa-
tions, and forgetting their own amiable character,
would vigorously contest for the possession of
their persons.

The only manufactures carried on in Fundah
are cotton cloths, extensive dye-works, and iron
and copper utensils. The cotton is spun by every
one, from the king downwards, upon what, I
believe, is called a falling bobbin. The spinner
carries a quantity of raw cotton in his hand, and
with his finger and thumb commences the thread,

the end of which he fastens to the bobbin, giving it a rotatory motion, and lets it fall, paying out the cotton with great quickness and dexterity. This is repeated until the bobbin reaches the ground, when the thread is wound up, and he begins again. I have frequently met slaves carrying large burthens on their heads, and spinning in this manner as they walked. The thread is rough but strong, the cotton being of very fine staple. The cloth is wove in pieces about twelve feet long, and three broad; is exceedingly durable and heavy : its price is about one thousand cowries per yard if plain, but, if dyed blue, about fifteen hundred.

They manufacture from native iron, hatchets, chisels, nails and clamps, and have a number of European locks and keys, particularly padlocks. Copper is used in ornamenting and fastening their large calabashes, and in the manufacture of bowls for their tobacco pipes. I tried in vain to get some of the ore, which must be very rich; but, though they readily made me several pipe-bowls to order, for five hundred cowries each, I never could get them to give me the ore or material of which they are made. The natives all agree, that it came down the Shary from the

eastward.   In all rude states of society the profession of the blacksmith is held in high estimation, and the Vulcans of Fundah are no exception to this rule; they rank next to the king
in importance, and are quite as consequential.

The trade of Fundah is trifling; the tyranny
of the king, and the number of years in which
the country has been in an unsettled state, are
sufficient reasons for this.   If the accounts of
the natives can be believed, it was formerly a
place of considerable trade—a sort of entrepôt
where the Arabs and Felatahs exchanged European goods for slaves.   Whether this tradition
be true or not, the town of Fundah bears inter-
nal evidence that it was founded by a race of
men superior to those who now inhabit it.   The
people who built the walls and dug the ditches
that surround the town must have been either
in themselves superior, or governed by a much
superior set of men to any I met there.

The subject which interested me most, and on
which I made many enquiries from every person
whom I thought likely to give me information, was
the course of the Shary.   The answers invariably
were, that it came from Lake Tehad; and one man,
a native of Kooka, offered to take me up there

in twelve days without changing the canoe. The
latter expression struck me forcibly, and I cross-
questioned the man closely; but he adhered to
his statement, and enumerated a long list of
towns on the banks of the river.   This evidence,
though strong, is by no means conclusive; the
natives are such inveterate liars, that no dependance can be placed on any accounts they may
give; and from the experience which I have had
of their mendacious propensities, I do not allow
their accounts to weigh for a moment against
the following reasons, which induce me to think
that it has its rise in another and very different
quarter.

The water of the Shary is colder than that of
the Niger.

The rise of the river commences sooner and
more suddenly than the Niger.

There is little trade upon the Shary in comparison with the Niger, which, if it communicated with the sea of Soudan, would naturally be
immense.

From the three first reasons, I should think
that its rise is in a mountainous country, and
that that country lies very near the equator.
Probably the same range of hills that gives birth

to the Cameroons, Malimba, and other large rivers, throws off, on its opposite declivity, the sources of the Shary. From its sudden rising it appears to me that it cannot be caused by the gradual and regular overflowing of an immense inland sea. This is the opinion I formed against the reports of the natives, and against the appearance of the country, as the range of hills that commences at the confluence diminishes its elevation gradually, and I have no doubt crosses out at no great distance. So there did not appear to me any physical difficulties that would prevent the river from trending to the northward. Future travellers will settle this question, and it may appear hereafter that I have been mistaken in my opinion. If so, I will most willingly acknowledge my mistake, as it will facilitate in a great measure the civilization of the people, by admitting commerce and its attendant blessings into the very centre of the country.

CHAPTER X.

Departure from Fundah.—Arrive at Yimmahah.—Return on board the Quorra.—Determination to return down the River to Fernando Po.—Description of Adassah and its King.—Sarsfield and Lieutenant Allen proceed to Fundah. —Ravages of the Felatahs.—Mortality among our Kroo-men.—Hector's Narrative of his Visit to Fundah.—Insubor-dination among the Kroomen.—Anchor off Bocqua.—Joined by Mr. Oldfield from the Columbine.—Letters from Europe. —Account of Mortality in the Columbine.

AFTER leaving the rabble that followed us out of the town, I proceeded in what my Kroomen insisted was the right road to Potingah, and, after a ride of two hours, was much annoyed to be told by a man who overtook us on horseback, driving a string of slaves before him to market, that we had taken the wrong path after passing the first ravine, and were considerably to the northward of our course. Being already much fatigued and rather startled at the information, I rode up to a farm-house, where I was welcomed by the women, and lay down in the shade for a short time, until the proprietor of the house came. I engaged him to conduct us to Potingah, and we set out again, my conductor holding me on my horse, as the sun was now very powerful and my weakness fast increasing. I was unable to proceed far, and in an hour's time was obliged to stop and be lifted off again at another farm-house, where I lay for some time almost in a state of insensibility.

In the evening I was aroused by Sarsfield, who had become alarmed at my not appearing at Potingah, and had found me out by my guide going and informing him of my situation. It appeared that he had had some difficulty in getting away from Fundah after I had left; and Smith the mulatto, who was ill of dysentery, refused to accompany him. He had left a bag of cowries with him, and the king promised that he would send him down, in a canoe from Potingah, as soon as he recovered. As the sun was now less powerful, and there was a pleasant breeze, I was mounted on my horse again, and, about eight o'clock in the evening, was once more afloat in the same abominable canoe which brought me from Yimmahah.

As I was very ill, I kept Sarsfield with me, directing the Kroomen to proceed by land to Yimmahah, and we went on in the canoe. At about eleven o'clock at night, one of the most awful tornadoes that I ever witnessed came on, and continued for four hours. The canoemen directly pulled alongside the bank, and speedily took themselves off into the bush, leaving Sarsfield and me to take care of ourselves and the canoe also. Fortunately we had some brandy, which, with a little opium, carried us through the night. Having no covering, excepting our blankets, we literally lay in the water; and if the canoe had not been aground, she must have sunk from the quantity of rain that poured into her. A number of fowls that we had tied to the thwarts were drowned.

At daybréak the clouds cleared away: we spread our clothes out to dry in the sun, and, clearing out our canoe as well as we could, proceeded onwards. After a roasting from the sun, as a sort of contrast to the soaking we had experienced, we reached Yimmahah in the evening, where I entered my own boat, and enjoyed the hurricane-house in her stern as much as if it had been a palace, which it really was, compared with those which we had been lately inhabiting.

On my arrival I was visited by my old friend the Mallam, and a number of my old acquaintances, who came to congratulate me on my return; all of them bringing presents of some trifling kind. I rewarded the old chief for taking care of my boat, not a rope-yarn having been stolen; and I may here remark, that the whole time I was at Yimmahah and Fundah, I never lost an article of the slightest value from theft by the natives : my own Kroomen pilfered and cheated me in every way; but the natives, although keen bargainers, are honest, or rather they were so to me; and even the king never took anything which I refused him, although he would threaten and bluster about it. My goods were quite at their mercy, if they had been inclined to take them ; and, considering the condition in which I was, their refraining from doing so may be looked on as a decisive proof of the facility and security with which trade may be carried on in the country.

In the morning, after a night's rest much needed by us all, we departed from Yimmahah ; and in the evening of the second day arrived on board

the Alburkah, where I was glad to find Lieutenant Allen perfectly recovered.   I received from
him a note, written to me by Mr. Lander, merely
telling me, that he had thought proper to take
Captain Hill to the sea, in order to give him a
chance for his life.   The next morning I went
down to the Quorra, and found her afloat, the
river having risen considerably since I left her.
I found Hector, Harvey, and the rest of the men
much in the same way, in point of health, as
when I left them; and, for my own part, I had
returned as weak and helpless, after two months,
as I was when I started.

As it appeared to me, that it was very impro-
bable that Mr. Lander would return : that in the
present state of my own health there was little
chance of my surviving if I remained in the river;
and as we had been long enough in communication with Rabbah, Egga, Toto, Domo, and other
towns, that had been represented to us to abound
in ivory and other productions, to discover that
the expectations which had been raised by those
representations respecting the profitable nature
of the trade were completely falsified, inasmuch
as we had not collected more than forty-five hundred weight of ivory, I determined to take the

vessels down, and lay them up at Fernando Po, confident that keeping them up the river would only add to the enormous expense that had already been incurred, without the slightest chance of remuneration; and being aware that, with the debilitated crews of the vessels, the only chance of safety consisted in taking them down on the rising water.

As I did not consider that the river had risen sufficiently to enable us to proceed without danger of stranding, I determined to pay Cuttum-Curaffee a visit, as I found that our tallow was expended, and I had been informed by the natives that plenty of Shea butter could be obtained there, which would answer equally well for the engine. Although not thirty miles distant, this town had not been visited by any person belonging to the vessels. The chief was indebted to us for a considerable quantity of goods, for which there appeared little chance of getting anything, without a personal application, and this circumstance formed an additional reason for my going to Cuttum-Curaffee at this time.

After a few days' rest, taking Mr. Hector and seven Kroomen, I set out for Cuttum-Curaffee, and slept the first night on board the Alburkah. Leaving her in the morning, I proceeded up the Niger, and found an excellent channel running N. E. by E. for five or six miles, then N. and afterwards N. E. The scenery is very beautiful on the western bank, the meandering of the river through the hills giving it a very picturesque appearance.

On the evening of the second day, we arrived at Adassah, a small town, about five miles from Cuttum-Curaffee, and the nearest point on the river. A creek, passable only for canoes in the flooded season, runs up to the town. Immediately on our arrival, we sent for horses, which arrived about twelve o'clock next morning. Leaving two men with the boat, we mounted our steeds which were but sorry creatures. Our path wound amidst plantations of grain and yams for a short distance, and then struck off due east, through a dense forest, crossing the creek, which took the horses up to their bellies. We found the remains of a bridge across it, consisting of upright posts, with forked heads, one or two of the horizontal pieces of which were still remaining. My conductors informed me, that it had been destroyed by the Felatahs, when they burnt the town. The banks are of clay, and very steep. Hector had a narrow escape : in attempting to get his horse up them, the animal tumbled backwards, and he was saved only by throwing himself off into the water.

After a ride, which was to me very distressing, we came in view of the town, prettily situated under a range of low hills. The place would, I think, be very healthy, were it not for a swamp on the western side. Through this swamp we followed the path, which was raised above its level, by a causeway three or four feet high. The town is walled and ditched, but presents a very ruinous appearance, not having recovered from the attack of the Felatahs, who sacked and burnt it twelve months before I visited it. The present inhabitants, in my opinion, do not exceed five thousand, though, from the extent of its walls, it may have contained four times that number. I passed through the town, entering at the western, and going out at the northern gate, as I was determined to sleep outside of the walls, amongst the Housa people, who have a village to themselves, beautifully seated on the side of the hill. I took possession of a very clean, neat hut, belonging to a woman with a large family. Her eldest, a girl of thirteen, was one of the most perfect models, as to figure, I had ever seen; her foot and ancle were exquisitely beautiful, and her carriage was at once graceful and dignified. She was unmarried, and wore merely a cloth around her loins, but was intended for the chief; I believe, however, that thirty thousand cowries, or about thirty shillings, would have bought her.

If ever the musquitoes had a paradise, it must have been at this town, for the myriads that greeted us the first night of our arrival was beyond anything I had yet witnessed even at Fundah, which I had thought so bad that nothing could be worse. They had long ceased feasting on me; but Hector, who had recovered wonderfully from his illness, and had some flesh on his bones, was too good a subject to be missed, and, when daylight arrived, I hardly knew him, so completely was his face besmeared with blood that had been oozing out of the numerous bites all night.

In the morning we paid a visit to the king, who was rather annoyed at our not staying inside the walls of the town. He was an old and infirm man; said he was very poor, which his habitation evinced, as it was completely in ruins, but had evidently been a very large place. I gave him a present, for which he promised me some bullocks. I was disappointed to hear that no butter was made nearer than Domo, which was represented to be distant three days' journey :—the king had no ivory, nor anything to trade with. I took my leave after a short audience, fully satisfied I should get nothing, and feeling pity for the old man, who evidently had seen better times.

The town presented an appearance as if it had been sacked, and half destroyed ; roofless houses and broken down walls were seen on every side, with the rank vegetation springing up among them in all directions. On passing the ditch, I observed several human skulls, and other bones. The Felatahs had assaulted the place upon this side, and, if the inhabitants can be believed, were repulsed three times before they stormed the town.

The exertion of the last two days, and my exposure to the sun, brought on an attack of dysentery, which confined me to the hut. Hector, who was well, rambled about the town, but saw nothing remarkable. I took my old friend the barber's prescription of raw rice and water, and

by confining myself to it alone, became well
enough in a few days to sit upon horseback. My
landlady was very attentive to me during my
sickness, and gave me great relief by dressing
my sores, as I had become so emaciated that my
hip and elbow bones had cut through my skin;
and the irritating disease I had contracted had
broken out into ulcers all over my body. I determined to leave Hector to await the arrival of
the messenger that had been despatched to Domo
for the Shea butter, and return to the Quorra;
sending the boat back for him immediately. The
king very civilly supplied us with horses; and
after a most distressing ride, in which, short as
it was, I fainted three times under the excruciating pains of my sores, I got into my boat.
Hector returned to the town, and the next evening I reached the Alburkah. On my telling Mr.
Allen that the boat would return for Hector, he
expressed a wish to visit the town; and on my
sending the boat back in a day or two with Sarsfield, he took a passage in her, and fixed the
latitude 8° 7′ N. and the meridional distance,
east of Stirling, at the confluence, 0′ 12″ E.

In a few days Hector returned; he had been
but partly successful, having got but very little

butter and no payment from the king. I had been much pleased with his conduct while at Cuttum-Curaffee; and as I could place the greatest dependance upon his coolness and self-possession in any situation, I determined to send him up to Fundah, to try and get some repayment for the quantity of things I had been cheated out of there, and which I attributed in a great measure to the infirm state of my health, which prevented my making sufficiently energetic exertions to recover them.

As the river was rising rapidly, and no time was to be lost, two days after his return I despatched him and Sarsfield with six Kroomen, with orders to make no stay if they did not succeed at once in getting payment. They were to call on board the Alburkah, Lieutenant Allen being desirous of visiting Fundah. On the 19th of June, they left us, and I heard of them two days afterwards by a canoe from the Shary, that had met them some distance up the river.

While at Cuttum-Curaffee, I heard that the Felatahs were expected to make a descent upon the western or Kacundah side of the river, and a few days after the departure of the boat, the whole of the inhabitants of the towns on that side of the river gave notice of their approach by flying in dismay to the opposite bank, which for many miles was covered with their barracoons, or temporary huts hastily erected of mats. Great numbers encamped close to us, particularly our Addah Kuddah friends, who had, at the first intimation of the approach of their dreaded enemies, conveyed all their little property in canoes to the eastern bank, putting the river between them and the Felatahs, who being destitute of boats could not follow them. It was amusing to hear the various reports of their strength, all of which differed, in proportion to the terror of the informer, from five to fifty thousand, a considerable part of which were said to be horsemen. A few days after the arrival of the fugitives, a column of smoke, rising in the air about five miles above the confluence, marked their advance, and in two days afterwards the whole of the towns, including Addah Kuddah and five or six others, were in a blaze. The scene at night was imposing; the fire, catching the dried grass, ran furiously along the ground, excited by a strong breeze from the westward, which rolled the dark mass of smoke over the river. The shrieks of the unfortunate wretches that had not escaped, answered by the loud wailings and lamentations of their friends and relations, (encamped on the opposite bank of the river,) at seeing them carried off into slavery and their habitations destroyed, produced a scene which, though common enough in the country, had seldom, if ever, before been witnessed by European eyes, and showed to me in a more striking light than I had hitherto beheld it, the horrors attendant upon slavery.

I had brought down the Alburkah from her anchorage at the confluence at the first notice of the approach of the Felatahs near my own vessel, having no one on board her that I could trust to, and not being aware of the intentions of the Felatahs towards us. The towns continued burning several days; but the invaders evidently did not wish to venture beyond our vessels, as none of the towns were fired below us. As I was exceedingly anxious to open a communication with them, and the natives having reported that they were encamped near Stirling, at the confluence of the rivers, I sent up Harvey with a boat's crew, with orders to land unarmed, and endeavour, if possible, to communicate with them, giving him at the same time a number of presents and several

Arabic circulars, which had been printed in England, setting forth our object in coming as traders.

Harvey landed at all the towns, but found them deserted; the roofs of all the houses were burnt and the clay walls only standing. He left the circulars as I had desired him, in a conspicuous place in each town. Two other attempts were made to open a communication with them afterwards, without success: the Arabic papers had been taken down, but there was no appearance of the Felatahs; they had in fact retreated as fast as they advanced, carrying off a number of slaves with them. I regretted much that I was so ill as to be unable to make the trial myself, though I believe it would have been equally unsuccessful, as we happened to be unfortunately situated,—the natives having given out that we were their allies, and their crowding round our vessels gave some colour to their report. I have little doubt that this prevented our meeting with this extraordinary people, who, whether it may be from their physical power and mental energy, or their form of government, seem to be overrunning the whole country, and who, ere long, will most probably reach the coast.

My Kroomen had hitherto been very healthy, but while I was at Fundah, old Paskoe, Mr. Lander's servant, died, with strong symptoms of being poisoned; and one of the Kroomen named Accah, a fine young man of about twenty-six years of age, had died in a similar manner. Another was ill with the same symptoms: I tried the effects of mercury upon him, and salivated him severely; he apparently recovered, but relapsed, and died in a horrible state. He also was a healthy, strong young man, under thirty : his name was " Frying Pan." Jack Sabe, my Kroo cook, and a valuable man, was also ill with the same symptoms, and for the first time my Kroomen began to look frightened, and to grumble about their long detention in the river. They had hitherto behaved very well, but I saw that one or two more deaths among them would have a very disagreeable effect, and an occurrence shortly afterwards took place which confirmed me in the opinion, that their ready obedience could not be counted on as heretofore.

After an absence of twelve days, Hector returned from Fundah, unsuccessful, as I expected, and gave me the following account of his expedition :—

" After leaving the Quorra, on the morning of the 19th June, I received Lieutenant Allen on board, at the mouth of the Shary, and entered that stream at ten A.M. The breadth of the river, at this season of the year, is about a mile and a half, but broken by sandbanks and small green islands into three or four channels. The southern bank is high and sprinkled with trees; the northern one is lower, with jungle here and there. We proceeded slowly against the stream, and at noon stopped at a sandbank near some fishermen's huts. Ibrahim, the interpreter, was sent on shore to a small village, and purchased some eggs and dried fish ; we then proceeded until dark, when we stopped under the bank, the Kroomen lighting a fire on the shore and sleeping round it.

" 20th. We proceeded at daylight through the same sort of country : at noon the river became contracted to a breadth of five hundred yards, with a strong current, which the Kroomen had great difficulty in stemming by tracking the boat : both banks being thickly covered with wood, they were obliged to wade up to the middle in water. At sunset we emerged from the woods, and entered a noble reach of

the river. To the south was an extensive green
plain, dotted with trees and bounded by gently
swelling hills, partially clothed with wood. We
saw no appearance of inhabitants on this beautiful plain, unless immense numbers of monkeys,
that seemed to have possession of it, might be
considered such. The Kroomen took their evening meal on a sandbank, after which we proceeded during the night, though with great difficulty, owing to the intricacy of the channels.
At nine o'clock, on the morning of the 21st, we
arrived at a village on the north bank. I admired particularly the situation of this place: it
was a little removed from the clay banks of the
river, which were about twenty-eight feet high,
and was almost hidden by trees. Behind it rose a
conical hill to the height of six or seven hundred
feet, partially wooded, but very steep and rocky.
The natives sold us some provisions, particularly
some hung and smoked beef, which I had not
procured elsewhere. It was the flesh of the
wild bullock, and very good. At noon we passed
an immense number of fishermen's huts: an air
of great comfort and plenty was apparent among
them. The inhabitants were all employed in
catching and preserving fish, while around the

huts swarms of children were playing. Many women, with ornaments of native copper and brass bracelets tastefully embossed, paid us a visit. At seven P. M. we stopped for the night off a small village on the south bank.

"22nd. At three o'clock this morning, a tremendous tornado came on : the boat dragged her anchor, and grounded on a sandbank, where I fully expected she would have gone to pieces. At daylight the tornado abated, and at ten o'clock we reached the town of Yimmahah.

" This is the first Fundah town ; it stands upon a basaltic rock, that abuts abruptly on the river : this rock has a kind of natural staircase up it, formed by projecting stones, and is very difficult of ascent. I paid a visit to the chief, who pressed me to stay, and seemed vexed at my quick departure. A respectable-looking man came and informed me, that if I desired to go to Fundah by land, his horse, the only one in the place, was at my service. I declined the offer, and made him a small present.

" On the land side of Yimmahah, the ground descends, and again rises towards the hills, which are a continuation of the chain we had before seen.

" 23rd. We proceeded onward, the river becoming very broad, with its principal channel on the north side. At sunset we stopped, to allow the people to cook their meal; and then, with a fair wind, went on during the night.

" 24th. Early in the morning we reached the mouth of the creek or river called Okna, which leads to Fundah, into which we entered.

" 25th. At noon reached Potingah, a village about ten miles from Fundah, and despatched a messenger to the king for horses. Early on the 26th, we removed our baggage into the chief's hut. At ten o'clock the horses arrived, and we set out. The way we took was a bridle-path, which crosses three ravines. On emerging from the last, we saw the walls of Fundah at the extremity of a small plain covered with verdure and interspersed with magnificent trees.

" After entering the town through a gateway, we proceeded down the principal avenue for about half a mile, as far as the walls of the king's house, where in an open space we found some of the people exercising their horses.

" On our arrival being reported to the king, his head eunuch made his appearance and proposed to show us a proper lodging: we followed him to

several miserable huts, to which we objected, and
at last got a comparatively decent one belonging
to a mallam. The king sent us a present of a
sheep, rice and yams, in the evening, and crowds
of people visited us.

"27th. This morning about noon the king,
attended by his mallams and eunuchs to the
number of forty or fifty, paid us a visit. He
was ushered in completely covered with a cloth,
which his people took off when he sat down, and
displayed to our view his majesty swelled up
with cushions or cotton to the size of a hogs-
head. After the usual compliments, he asked
after Mr. Laird—if he was alive, and what we
wanted. I informed him, that we were here to
trade with him and his people, and to demand
the goods left by Mr. Laird, or payment for
them.

"His majesty did not seem very well pleased
at this demand, and said he merely came to com-
pliment us, and not to talk on business. Lieutenant Allen made him a present of some toys,
which he received very coolly, and shortly after
took his departure.

"He is a stout, ill-looking man, about forty-five
years of age, and has red eyes, which our Kroo-

men affirm is a sure sign of a bad man. In the evening he came back, naked, except a waist-cloth, and stayed a few minutes ; but Ibrahim the interpreter being out of the way, we could not make out what he wanted.

" 28th. It being a holiday to-day, we accompanied the king to a plain, on the east side of the town, where we witnessed several horse-races and native dances, which were performed before him to the noise of drums, trumpets, and other instruments. On returning with the king, he recognised in the market-place a native of Bassa, and a great squabble of tongues commenced. The man was alone, but retorted the abuse with great spirit. The people of Bassa and Fundah are often at war, and perpetually harassing each other. It was annoying to find, through our interpreter, that the king was threatening the Bassa people with the vengeance of the white men, who would, he said, utterly destroy their country.

" 29th. I was witness to-day to a poisoning scene, which it would appear was a favourite punishment at Fundah. The culprits were two women, who were placed under a tree in a court-yard, and an old man beat up the leaves of some herbs in a sort of mortar, mixing them with water, the women sitting quietly looking on. The liquid, which was of a greenish colour, was poured into two calabashes, and the women drank it off without any apparent reluctance. They then commenced walking quickly up and down the court, drinking large quantities of water from a calabash placed in the centre of it. In about half an hour, they both began to stagger, and totter in their walk; and in a few minutes more the tragedy was ended by their falling flat on the ground and expiring in apparently dreadful agonies.   The natives seemed accustomed to such scenes, and stated that it was the punishment for stealing the king's yams.

" In the evening I had another interview with the king, and found it impossible to procure either the goods or payment; indeed, I was abused and threatened; and knowing that my return would be anxiously looked for by Mr. Laird, I determined to leave in the morning, and demanded horses from the king for that purpose. His majesty, however, would neither lend us horses, nor suffer us to hire them; and on the 30th we departed on foot for our boat.

" We passed through the city, followed by a

great number of people; and on our arrival outside the walls, found that our interpreter, Ibrahim, had absconded with his musket and accou-
trements. He was a native of Kano, an intelligent and active fellow, but a great rascal.

" Our march to Potingah was not the most
agreeable, as we found the water in the ravines
in some places up to our waists, and we felt
really thankful once more to get on board the
launch.

" On the 2nd of July we arrived on board the
Quorra, and found the Alburkah at anchor beside her; having met with nothing but kindness
and civility from the common people, and roguery
and insult from their chief. In fact, I do not
believe, that if the king could have mustered
courage to attack us, his people would have supported him in his intention. "

After Mr. Hector's return, as I did not consider the river, although much swollen, was sufficiently so to justify our proceeding to the sea,
I employed him in trading with the launch to
Bocqua market; and a circumstance occurred to
him, while on one of his trips, strongly illustra-
tive of the disorganised state of the country, and
the demoralisation of the people.

On our voyage up, King Obie had presented
Mr. Lander with a slave, a native of Kacundah.
While the Alburkah was lying at the confluence,
he absconded, and nothing was heard of him until he was discovered by Mr. Hector, at Bocqua,
where he was on sale; his own brother having
enticed him on shore from the Alburkah and
brought him down to the market as a slave. On
seeing the boat, he made a run for her, and getting hold of the gunwale, roared out for assist-
tance. His brother claimed him as his slave and
seized upon him; but after a violent struggle he
managed to get into the boat, and as he was then
under the British flag, he was carried off in triumph, glad to get back to his old masters. He
was remarkably stout when he absconded; but
we hardly knew him when he was brought back,
he was so dreadfully emaciated. Several hundred people had crowded to the boat, shouting
and encouraging the man to assert his right to
the slave; and matters appeared so alarming, that
Hector ordered the Kroomen to fix their bayonets on their muskets, while he primed the swivel
and pointed it at the crowd. At these threatening movements they disappeared in a few seconds.
As the man was carried off in the face of the

whole market, consisting of five or six thousand people, by a boat's crew of two Europeans and six Kroomen, it affords a proof of the security with which a European may traverse the river in this part.

On Hector's last trip to Bocqua, a circumstance occurred which fixed my determination to get down the river as quickly as possible. My head Krooman, Tom Kay, so far forgot himself as to strike Mr. Hector and refuse to work at the oar. This was the first time that the Kroomen had shown the slightest symptom of insubordination, and coming from their headman, it became more alarming. When it was reported to me on the return of the boat, I determined at all hazards to put the man in irons; and in the morning, having sent the rest of the Kroomen ashore to cut wood, called him up, and charged him with his offence, which he did not deny, but boasted of it. Harvey and Hector immediately threw themselves on him, and got him down and put his hands in irons. On releasing him, he was furious, and attempted to jump overboard, but was prevented by two of his boys that were on board. On the remainder coming on board to dinner and finding their chief in irons, they all came aft upon the poop, where I was lying upon the skylight, propped up with pillows. Hector, Harvey, and Sarsfield, being the only men able to stand, were beside me well armed, and the two field-pieces loaded with grape were pointed forward. They were going to speak, when I ordered them all off the poop except "After Dinner," the second headman, threatening to fire upon them if they did not obey. After a moment's hesitation, they went off; and I acquainted Mr. "After Dinner" with the reason why I had put his chief in irons, and told him he was to take charge of the Kroomen from that time. I then called the other men, and informed them that Tom Kay was no longer headman, and that "After Dinner" was to be considered as such. They obeyed quietly, but the peculiar character of the people now evinced itself. I had disgraced their headman, and was obeyed from fear and with reluctance ever afterwards. While he remained in irons, their attention to him was beyond anything I could have imagined. They gave him the tit-bits of their mess, and made up, out of their own, his allowance of grog, which was stopped. Though disgraced, he was their headman still, and was treated by them amongst themselves with more respect and deference than before; though, as far as the duty of the ship went, they obeyed " After Dinner" in everything. This trait in their character gave me a higher opinion of them than I hitherto had, as it contrasted so forcibly with the treatment which a petty officer receives when he has the misfortune to be broke amongst more civilised people: he may be pitied by his superiors, but he is always laughed at and jeered by the class he is lowered to.

On the 10th of July, the water having risen by measurement fourteen feet, we got under weigh, taking the Alburkah in tow, and proceeded rapidly down the river, passing the beautiful gorge through the Kong Mountains, and arriving at Bocqua, where we anchored to purchase rice and other provisions. About two o'clock in the afternoon, to my great surprise, a boat under canvass hove in sight. I sent a boat to her with Hector, who returned with Mr. Lander and Mr. Oldfield, our surgeon, bringing a parcel of letters for me. Those only who have been shut out from all communication with friends and the civilised world for some time, can imagine the true value of a letter. I had left Liverpool when the

cholera was exciting so much alarm and commit-
ting such ravages, and had almost feared to hope
that my friends had escaped.   How gratefully
then did I return thanks to the Giver of all
good, that none of those that were near and dear
to me had fallen !

After visiting his friend Mr. Allen, Mr. Lander came on board with Mr. Oldfield.   He was
looking much worse than when I last saw him,
and had evidently suffered much in the boats
from exposure to the weather.   He had been
thirty-two days on his passage from the Nun, in
the brig's long boat, which he had thatched over
abaft, and had brought up with him a Mr. Dean
to command the Alburkah, and one white man
as a sailor.   Mr. Oldfield told me that the mortality in the brig had been very slight in comparison with other vessels that had entered the
river, and he attributed the circumstance to my
having anchored her open to the sea-breeze.   I
learnt from him that an American expedition
had arrived in the river, consisting of a brig and
two small schooners.   The supercargo of the
brig having been unfortunately killed by the
bursting of his gun, the vessels had been obliged
to return without accomplishing their object, and

after having lost most of their men. The only
other vessel that had entered the river, was a
Spanish slaver from the Havannah, that in a
few days lost fifteen of the crew. Mr. Oldfield
himself had suffered from fever ; but he had recovered, and was now in good health and looking well, and from the severe seasoning he had
undergone, would, I had no. doubt, remain so.
At the mouth of the river his time had been
passed most monotonously, and from his description, even our own situation must have been
far more preferable. Captain Mitchell and two
seamen had fallen victims to the climate, and the
others had all suffered more or less from fever.
Leaving me to peruse my letters, they returned
on board the Alburkah. Mr. Lander had the
kindness to send me two bottles of Port wine,
one gallon of brandy, and some tea and flour,
which he had brought up the river. Accompanied by Lieutenant Allen, Mr. Lander and Mr.
Oldfield came on board to dinner, and with true
professional delight the latter set to work upon
me. He at once pronounced my complaint to
be an inveterate itch, aggravated by the excessive debility of the whole system; and commenced his operations by dressing the ulcers

with which I was covered, and plastering me over with brimstone ointment every day, afterwards steaming me in a portable steam-bath that I very fortunately had brought out with me.

CHAPTER XI.

IN a few days I felt considerably relieved by Mr. Oldfield's treatment; and on the 10th we departed from Bocqua, and in three hours arrived at Attah. The Alburkah followed, with Mr. Lander. In coming down, we took the inner channel, passing close under the perpendicular face of the hill on which Attah stands, and came to anchor a little below the landing-place. The river must have risen since we passed on our voyage up more than thirty feet, as we had four fathoms where it was impassable for boats in December.

On the 18th, my cook, " Jack Sabe," a Krooman, died, with similar symptoms to those that appeared upon the two others; viz. a burning pain in the pit of the stomach, excessive thirst and debility, with swellings of the body and extremities. Mr. Oldfield examined the body, but by no inducement could the Kroomen be persuaded to allow it to be opened. I have no doubt in my own mind that the deaths of these three Kroomen, and that of old Pascoe, were occasioned by poison, administered to them by the natives. The body, immediately after death, becomes putrid, and seems ready to burst, as a dark-coloured offensive fluid oozed from the mouth and nose.

Mr. Lander having determined to endeavour to reach Boussa in the Alburkah, fixed his departure for the 27th; and though very anxious myself to visit a place hallowed by the melancholy fate of Mungo Park, I did not consider myself justified in returning, and, for the reasons before mentioned, adhered to my determination to lay the vessel up at Fernando Po and return to England. On my informing Mr. Lander of this, he strongly recommended Prince's Island in preference, from many advantages it possessed over Fernando Po. Mr. Lander also informed me, to my great surprise, that he had added to our flotilla, by purchasing a schooner for the expedition, in which he was desirous I should take my passage home, as he represented her as a vessel calculated to make a quick voyage.

I was very sorry to part with Mr. Oldfield, who had been of essential service to me during the fortnight he was with me, my sores having healed rapidly under his treatment, and the cutaneous disease appearing to be eradicated. Dropsy had, however, come on, and I was already very much swelled. It seemed to be fated that I should not recover from one attack before another was ready to commence upon me.

The Alburkah left us on her voyage to Boussa on the morning of the 27th; and I intended to have started at the same time, but was anxious to obtain one of the small country horses to take down to the coast, and one or two bullocks, which run wild in the bush here, and though small, are the fattest and best I have met with. I sent Mr. Hector on shore to the king to ask permission to shoot one, which he immediately granted, with the proviso that he was not to shoot a female one. Unfortunately the Kroomen brought down a cow, which gave great offence and spoilt my market for the horses.

On the 28th I got under weigh, and in a few hours was off Damuggoo, going down at the rate of ten or twelve miles an hour. When immediately opposite the village, a sandbank brought us up with such a shock, that I thought the vessel must have stove her bows in. I was lying upon the skylight, and was pitched off upon the deck. Harvey, who was on the paddle-box conning the vessel, saved himself by clinging to the fore-topmast backstay. The man at the wheel and all upon deck were laid prostrate. On getting the boat out, we found ourselves upon a bank, with only one foot water upon it; and the vessel had run up so high, that it was impossible to get her off without waiting for the water rising, or the bank washing away. It was a melancholy spot to remain in, for immediately opposite we had deposited the remains of nine of our companions on our voyage up.

We continued, nevertheless, aground for thirteen days, when the gradual rise of the river floated us over the bank into nine fathoms water. The morning afterwards we got under weigh, and came to anchor off the town of Eboe on the evening of the 9th of August.

In the morning we were visited by a great number of the inhabitants, and, amongst the rest, by our fat female friend, who was overjoyed to see us, and brought a supply of palm-wine and bananas as a present. Obie sent his two sons on board to offer us anything we wanted, and to invite me to visit him. As I was by no means well, and still unable to walk, I sent Hector, whom he received very kindly, but was disappointed I did not visit him. He promised to send one thousand yams and twenty canoes loaded with wood on board the vessel in the morning, provided I would go ashore and visit him, as he wished to speak to me about the palm-oil trade.

In the morning I went ashore, taking Mr. Hector with me, and was carried up to Obie's house, who received me with great kindness. The palaver began by his complaining, that we traded with every person but him; that he was the greatest palm-oil king, had command of the river, and would not allow us to pass in future unless we bought and sold with him. To this very sensible remark, I replied, that we were not aware of his having palm-oil in such plenty, and did not know what goods he wished in return. He enumerated the following, and their rotation will show the value he set upon them :—

1st, Cowries.

2nd, Red cloth.

3rd, Red beads (mock coral).

4th, Soldiers' jackets.

5th, Romals and Bandana handkerchiefs.

6th, Rum.

7th, Muskets and powder.

This may be considered a list of the goods most suitable for the Eboe market; and I was surprised to see that he placed as the last in his list, guns, powder, and rum. He expressed himself with great contempt respecting the Brass and Bonny people, whom he characterised as rogues and vagabonds, that were obliged to come to his country to get their provisions, and said, if we would only promise to trade direct with him, he would send his two sons to our country to be made to know white men's palaver. His second son is a remarkably handsome, intelligent lad of sixteen, and I regretted that it was out of my power to take him. He had lent Mr. Lander four boys to assist him in working his boat up the river: I had brought these down with me, and upon examining their things when leaving the vessel, found that two of them had secreted a number of articles in their mats. Of course I detained them, intending to flog them. I now told Obie of their conduct and my intention, and requested him to send his sons to witness the punishment. He made some objections to my flogging his people. I persisted in my intention, as they had been caught in the fact, and in our present situation the slightest symptom of weakness or indecision would have ruined us. I asked him to give me the two other boys, one of whom was acting as my servant, which he readily acceded to, saying I might have a dozen more if I liked.

On taking my leave, he accompanied me to the water-side, and I returned on board completely exhausted. During my absence a great quantity of wood had been brought aboard, and I found that we most probably should be ready to start the next night.

In the morning the two young princes came off to witness the punishment of the boys. They were accompanied by a great number of canoes, and I must confess I felt rather nervous about

the result. However, it all passed over very
quietly ; and when the boys were cast off, they
were greeted by a general laugh from the
people, who had been looking on.   Three
canoes loaded with most beautiful yams came
alongside as a present from Obie, and another
with some goats and a small bullock.   In return,
I sent him a present of a variety of things, and
in the morning afterwards, having received on
board sufficient wood to take the vessel to the
mouth of the river, departed from Eboe.   If I
had been pleased with Obie's character before,
I was much more so now.   I had been completely in his power : the vessel's decks were crowded
with his people; they were aware that out of the
five white men I had living, three were confined
to their hammocks ; and yet I was received with
more kindness, and had more respect paid to
me, than when I visited the place before, with all
my crew living, and in full health and strength.

The only incident that is worth relating on our
passage from Eboe to the sea, was our mistaking
the Warree for the Nun branch, and proceeding
some miles down it before finding out our mistake. We had six or seven fathoms water in it ;
but where we turned, it appeared to spread out

to a considerable extent, which in alluvial soil is a sure sign of shallow water.

On the 19th of August we saw once more the salt water, and came to anchor astern of the brig Columbine. Coming as we did from the interior, we felt the sea-breeze piercingly cold, and were glad in the evening to take up our old quarters in the cabin.

I found the Columbine in charge of a stranger; this person and one of her apprentices were the only white people on board; the remainder had gone in the schooner on a cruise to Fernando Po. After waiting until the 26th in hopes of the crew of the brig returning, I got under weigh with the Quorra. I took the brig in tow and attempted to cross the bar with her, when, about half-way between the entrance of the river and the bar, a squall from the sea prevented our progress. The anchors of both vessels were let go instantly, and we brought up within fifty yards of the breakers on the eastern reef. While the squall lasted, the engine was kept working, and was one means of preventing our dragging.

As it was too late to attempt the bar again on that afternoon, we remained there riding very heavily. In the evening the schooner came in with the brig's crew, who were not a little surprised to see their vessel unmoored and half-way across the bar. With the morning's tide we again got under weigh, and succeeded in crossing the bar, on which we found eighteen feet at one quarter ebb: the sea upon it was terrific, the brig pitching jib-boom under.

On the 28th I saw the high land of Fernando Po, and getting up steam in the Quorra, ran into Clarence Cove at three o'clock the next morning, the brig and schooner following the same day in the afternoon.

On landing at Fernando Po, I was most hospitably received by Colonel Nicolls, who insisted on my taking up my abode with him during my stay in the island, and to whose judicious care and kindness I am indebted in a great measure for my complete restoration to health.

Two Liverpool vessels were in the bay, and the William Harris, transport, whose commander, Mr. Terry, was recovering from a severe fever he had been attacked with at Sierra Leone. On board one of the Liverpool ships, the Richard Rimmer, I unexpectedly met an old acquaintance, Mr. Jeffrey, who looked as well as if he had been direct from England, though he had

been several months at Bonny. His surprise at
meeting me was equally great, as the general
opinion was that we were all dead on board the
Quorra: indeed the variety of reports in cir-
culation (some of them not the most flattering to
my vanity) at Fernando Po was quite ridiculous; and I found, as I suspect most people in the
same situation will find, that every reason except
the true one was assigned as the cause of the
failure of the expedition.

By Mr. Jeffrey, who sailed the next morning,
I communicated to my friends in Liverpool my
arrival, and my intention to procure a charter
for the Columbine and bring her home myself,
if I did not succeed in finding a commander for
her,—informing them at the same time of the
complete failure of the expedition as a mercantile speculation.

Three days afterwards, I ran the Quorra
over to Calebar, accompanied by Colonel Nicolls,
who was desirous of a personal interview with
Duke Ephraim, on the subject of the slave-
trade. Calebar has been so long frequented
by British vessels, and so ably described by
Lieutenant Holman, that a description of it

now would be superfluous. I may remark, however, that I was much struck by the extreme demoralisation and barbarism of the inhabitants,
in comparison with the natives of the interior.
The human skulls that are seen in every direction, and that are actually kicking about the
streets, attest the depravity of feeling among the
people, and add another to the long list of melancholy proofs of the debasing effects of European intercourse with savage nations, when governed solely by the love of gain.

Colonel Nicolls entertained Duke Ephraim
and a great number of his followers at breakfast
on board the Quorra, and, with the exception of
the duke and three or four of his headmen, a
more disgusting and swinish assemblage I never
met with: far inferior were they to the Eboes,
and not to be compared with the decent and
well-behaved inhabitants of Addah Kuddah, and
other towns in the interior.

The duke's idol was brought on board,—a little
abominable figure of clay: the bearer of it followed him everywhere, and stood behind him
during breakfast. As the Quorra was the first
steam-vessel that had ever been at Calebar, we

made an excursion up the river for ten or twelve miles, to the great delight of our guests and the inhabitants.

There were three Liverpool vessels lying opposite the town, loading with palm-oil and red wood, of which articles this port annually exports from four to five thousand tons of the former, and great quantities of the latter.

After remaining three days, we returned to Fernando Po, which is distant from the mouth of the river about fifty miles; Duke's Town, where the ships lay to take in their cargoes, being sixty miles inland. From the entrance, or Tom Shot's Point, as it is called, to Duke's Town, it is a complete swamp, overgrown with mangroves, by which the vessels are shut in, the sea-breeze passing over it, imbibes the malaria arising from it, and causes the great mortality that distinguishes this port even upon this deadly coast.

On my return to Fernando Po, I recovered rapidly, and was able to walk and ride about in a fortnight after my arrival. The splendid scenery that distinguishes this beautiful island is well known from former descriptions, and to persons coming from the low marshy shore of the main land has indescribable charms. The view from

the galleries of the Government House on a
clear moonlight night, I never saw equalled, nor
can I conceive it surpassed. To the north-east,
the lofty peak of the Cameroons, rising to the immense height of fourteen thousand feet, throws
its gigantic shadow half-way across the narrow
strait that separates the island from the main
land; while the numerous little promontories and
beautiful coves that grace the shores of Goderich
Bay, throw light and shadow so exquisitely upon
the water, that one almost can imagine it a fairy
land. On the west, the spectator looks down
almost perpendicularly on the vessels in Clarence
Cove, which is a natural basin, surrounded by
cliffs of the most romantic shape, and a group of
little islands which Nature seems to have thrown
in to give a finish to the scene. Looking inland
towards the island, the peak is seen covered with
wood to the summit, with its sides furrowed with
deep ravines, and here and there a patch of
cleared land showing like a white spot in the
moonlight.

There is something inexpressibly soothing in a
serene moonlight night in the tropical climates :
the heat, fever, and agitation of the day are at an
end; all Nature is hushed into quiet repose, and

sheds an influence over the mind which accords well with her own peculiar stillness and composure. If she is beautiful in her calmness, she is magnificent in her wrath. Often have I stood watching the gradual appearance of a tornado : its mutterings and growlings in the distance contrasting so forcibly with the deadly stillness that pervades the atmosphere immediately around— the immense cloud which like a funeral pall covers the eastern horizon, occasionally illuminated by flashes of the most intense and dazzling lightning, which only makes the darkness darker — the rush and roar of its approach, and the peals that accompany it, altogether form a picture that it is impossible to describe.

I accompanied Colonel Nicolls to Cameroons, who in his judicial capacity had to inquire into some circumstances that had lately occurred there, which, as they show the lawless state of the coast, and the necessity for some resident British power to curb it, I may be excused for mentioning here. A short time before I reached Fernando Po, Commander Trotter, of his Majesty's brig Curlew, called at Prince's Island, and there obtained information that a very suspicious vessel, under Spanish colours, had been in Port

Antonio, and had gone over to the river Nazareth, on the main land, to load slaves. From the freedom with which her crew had been spending money, it was suspected she would prove to be a pirate, (as indeed slavers generally on the outward voyage make very little distinction between *meum* and *tuum*). He very promptly followed her to Nazareth; but his vessel drawing too much water for the bar of that river, he was obliged to enter it in his boats. On his approach, the crew deserted the schooner, leaving a lighted match, which happily was extinguished before communicating with the powder by Captain Trotter and his party. She was afterwards accidentally blown up, and some lives were lost by the event. The king of that part of the country, with whom the Spanish captain and crew had taken refuge on shore, refused to deliver them up, and Captain Trotter not being able to force him, returned unsuccessful to Fernando Po.

While he was there, Mr. Becroft, who has a commercial establishment at Fernando Po, and another at Bimbia, on the main land, went over to the latter place in a small schooner; and while there, nine Spanish seamen arrived in a little boat, worn out by fatigues and hardships. They informed Mr. B. that they had belonged to the Negro, a Spanish slaver from the Havana, that had come from the southward, where they had been wrecked, and they wished to get to Old Calebar or Bonny to join some of their country-men.

On his telling them he would take them to Fernando Po, and from thence they might get to Bonny, they hesitated a little; but as their boat had been stove in landing at Bimbia, they had no alternative. On arriving at the island, they were strictly examined by Colonel Nicolls and Com-mander Trotter; but all adhered to the same story, that they were shipwrecked and in distress. Colonel Nicolls seeing that one had a less har-dened look than the rest, took him aside and charged him with belonging to the Panda, the piratical schooner that had been blown up in the Nazareth. The fellow dropped on his knees and confessed they were part of the crew, and gave an account of the different vessels they had plundered since leaving St. Thomas's. On being confronted with the remainder, with the excep-tion of the boatswain, they all confirmed his statement, and a few days after my arrival this man changed his story and made a full confession. Instead of proceeding to Bonny, they were forwarded at the king's expense by the transport to Ascension. I must confess that my romance never had such a fall as when I saw these pirates; I expecting a fine-looking, dashing, daring set of men;—instead of which, they were a parcel of the most mean-looking scoundrels I had ever seen, and my sympathy for those they had plundered gave way to surprise at their having made no resistance to such a set of scarecrows. They reported that they had thrown away a considerable number of dollars at Bimbia, at their landing, to prevent suspicion in case they had been found upon them,—and to recover these dollars was one object of our visit.

We sailed at midnight, and in three hours came to anchor off Bimbia. At daylight we ran inside the island of that name, and came to anchor between it and the main land. We landed after breakfast, and were received by Prince William of Bimbia, who was rigged out for the occasion in scarlet trousers, fustian shooting-jacket, and a gold-laced hat. He described the landing of the pirates, the swamping of their boat, and paid into Colonel Nicolls's hands the dollars that had been fished up by his people.

He then introduced us to a dozen of his wives, and accepted Colonel Nicolls's invitation to accompany us back to Fernando Po. We re-embarked immediately, and proceeding a short distance up Bimbia river, we struck off to the eastward, through a narrow mangrove creek, that communicates with the river of Cameroons. In common with similar creeks, it had a bar at both ends, with only seven or eight feet water over them, inside of which there were four or five fathoms. This creek is about seven miles in length, and not more in any part than one hundred yards wide.

We entered the great basin of Cameroons river in the afternoon, which appeared about nine miles wide; and after visiting the Doddingstons, a large Liverpool ship lying ready for sea near the entrance of the river, we came to anchor in the evening, between King Bell's and King Agua's towns, and received a visit from both these potentates immediately on our arrival. The only vessel we found at the towns was the Amelia* of Jersey, trading with the natives for palm-oil and ivory.

* The melancholy fate of her commander and the attempt to condemn two innocent Kroomen are well known.

In the morning we went ashore to visit King Agua, who received us on the beach. The most striking article of his attire was an enormous cocked-hat, evidently "made to order," from its dimensions, and stuck round with small circular looking-glasses, which had a most dazzling effect as he bowed and scraped our welcome.

After viewing his house, which was of two stories, with a gallery surrounding it outside, we walked through the town, which in order and beauty far exceeded any thing I had yet seen in Africa. It appeared to me the *beau ideal* of an Indian town. The principal street is about three quarters of a mile in length, about forty yards wide, perfectly straight, and the houses being all upon the same plan, give it a regular and handsome appearance. In the course of our ramble we paid a visit to a person whose native impudence, aided by my countrymen's gullibility, had rendered him notorious. We found him reclining in his hut, attended by three of his women, who were employed dressing a loathsome sore upon his thigh. This man is a slave to King Agua, and acts as pilot for vessels coming into or going out of the river. In exercising his calling on board a Spanish slaver, he was carried off with his canoemen,—a method of proceeding by no means rare. The vessel was captured off the Havana by one of our cruisers, and he represented himself as Prince William of Bimbia. This story was so plausibly told, and he won the good feelings of the officers to such a degree, that he not only escaped the apprenticeship (as it is called, by way of burlesque, I suppose) to a Spanish master, but was favoured with a passage in the gun-room to Portsmouth. From thence he went to London, and was a lion for a short time,—was converted, baptized, and, after making promises to his patrons of his influence if ever they sent vessels to trade in his dominions, took a passage in the William Harris, transport, to Fernando Po, promising her commander a handsome remuneration in ivory upon his arrival. The imposture was of course immediately discovered; and he now figures in his old capacity as pilot, and is only to be distinguished from his untravelled countrymen by an increase to his original stock of roguery and vice.

On our return from our walk, I was gratified by an exhibition of maternal affection, which, I

regret to say, is but little known in this country.
Colonel Nicolls had brought with him one of his
boys, named Peter, a remarkably smart, active
lad, a recaptured slave. It seemed Peter was
a Cameroons boy, which we were not aware of
until our visit, when he was recognised by his
mother and claimed. The story of Peter being
carried off into slavery was exceedingly interesting to us all, and as I am afraid it is a common
one in this unhappy country, I relate it, to show
that the evils of the slave-trade are not confined
to the enslaved. Peter's father was a respectable
trader in Cameroons, and, amongst other trans-
actions, bought a canoe from a Bimbia trader.
He paid half the price down for this canoe, and
gave his son Peter as a pawn (as it is termed
upon the coast) for the other half. A slave-
vessel came into Bimbia river, and Peter's new
master was actively engaged in supplying her
with slaves. On the day of her sailing, he was
to bring on board a quantity of fresh provisions,
for which he had received goods in payment be-
forehand, and left Peter as pawn with the slaver
until he had fulfilled his contract. This he did
as the slave-vessel was getting under weigh, and
demanded his pawn; but the commander, having

got clear of the river, refused to give him up, and carried him off. The vessel fortunately was captured, and condemned at Sierra Leone. Peter was put to school there, and afterwards was taken by Colonel Nicolls as his servant—had accompanied him by mere accident to Cameroons, and was recognised by his mother, after five years' absence. His friends during this time had not been idle. The news of his being carried off soon reached them ; they commenced a partisan warfare, in which Peter's father and several others had already lost their lives, and numbers had been made slaves.

Peter was offered by his master, the Colonel, the option of returning to Fernando Po, or remaining with his mother; and he wisely preferred the security of the English settlement.

After leaving King Agua's town, we paid a visit to King Bell's, which is not so neat or clean as the other. It is situated on the same bank of the river, about a mile distant.

Another royal personage resides about three miles up the river, but we did not pay him a visit.

The two chiefs accompanied us on board to witness a race between a canoe paddled by

sixty-five men, and Colonel Nicolls's Deal galley
paddled by twenty-two Kroomen. The canoe
was much larger than any I had seen amongst
the Eboes, and measured above sixty-five feet in
length;— an immense length for the single tree
of which she was composed, and that tree hard
wood. Notwithstanding the exertions of her
crew, she was easily beaten by the galley, to the
great astonishment of the natives, and triumph
of our Kroomen.

We departed on the succeeding morning, saluting the towns, and, calling for letters on board
the Doddingstons, went by the outside passage
to Bimbia, where we arrived in the afternoon.
In coming out of Cameroons I saw the remains
of the Susan. This vessel was in the Nun when
I entered the river: she had been abandoned by
the crew off Prince's, and the current had carried her into Cameroons river, — a convincing
proof that there existed no real necessity for her
being abandoned. On the main land of Cameroons were the remains of another English vessel,
which it must have required no common ingenuity to lose in such a place.

On our arrival at Bimbia, we gave judgment
in an appeal-case relative to the death of a slave

The owner of the slave was his own counsel, and from his description there never had been, nor would be such another. The defendant replied in an animated strain, and after listening for two hours we decided in favour of the plaintiff, reducing his demand, however, considerably. The English language, every person knows, is capable of an infinite variety of dialects; but no amateur, however well acquainted with the King's English as spoken by the Lancashire, Somersetshire, or Yorkshire peasantry,— or, what is much worse, with its clippings and curtailings by the cockneys, — can imagine the variety it is capable of, without visiting the coast of Africa. In Bonny, everything is " a good or bad palaver ;" in another place it is " a good or bad go ;" and in Cameroons, everything is " a good or bad bob." Our " bob," unlike most " bobs," was highly satisfactory ; and we had all the gratification of preventing a feud between parties that, like Peter's, might have cost the life and freedom of many.

In the morning, Prince William and two of his wives came on board, as Colonel Nicolls had invited them to pay him a visit. The youngest was the favourite, as she was the last, I presume ; but the other was a much handsomer woman, and was really beautifully tatooed, which I do not dislike on a black ground, as in the absence of clothing it gives a finish to the skin. Through their ears and the cartilage of their noses, these ladies had pieces of ivory inserted about nine inches long; or, as the sailors called them, they had their " sprit-sail-yards and studding-sail-booms rigged out." I cannot say much for their feminine tastes, as they preferred rum to tea, and salt pork to the wing of a chicken! Their hair was dressed in a helmet-fashion, and looked very becoming,—even elegant: as to their bracelets, necklaces, and leglets, it would require an amateur of fashion to describe them.

On leaving Bimbia we ran along the coast as far as the entrance to the Rio del Rey, passing inside the beautiful little islands called the Amboisas, which form a secure and sheltered harbour for vessels blown off the mouth of the Cameroons river by tornados, and arrived at Clarence Cove in the afternoon.

CHAPTER XII.

Proceed to Calebar. — The French Captain. — Barbarous
Practices on the Death of a Chief. — Native Wrestlers. —
Return to Fernando Po. — A Slaver captured. — Fer-
nando Po. — Its advantageous Situation and Climate con-
sidered. — Natives and their Habits. — Voyage to Liver-
pool.—Arrival.

As I had discovered that some of the copper
sheathing had been rubbed off the Quorra's
bottom, I was obliged to pay another visit to
Calebar, there not being rise and fall of tide sufficient at Clarence Cove to lay her aground.

On my arrival there, I found a brig under
Brazilian colours, commanded by M. Gaspar, a
French gentleman, who had settled at Bahia, and
had come out in his vessel to lay in a cargo of
dyewood. As taking dyewood to the Brazils appeared to me very like taking " coals to New-
castle," I took it for granted that he had a very
different object in view, and, instead of dyewood,

was collecting a cargo of slaves for the Brazil market,—or, in the parlance of the coast, buying " ebony." The brig he had arrived in was an old crazy hulk, and certainly had not any thing suspicious about her. As to Gaspar, he was a most agreeable fellow, had served under Napoleon, was a survivor of the Russian campaign, and made a most entertaining and amusing companion. His officers were Portuguese; and a more unprepossessing-looking set I have seldom seen, contrasting strongly with the gentlemanly manner and appearance of their commander. His ridicule of British cruizers and his stories of his numerous escapes from them were related with infinite glee; and if one could have forgotten the character of the speaker and nature of his trade, he would have tempted any man to make a voyage with him.

I witnessed a most barbarous dance here, in honour of a chief who had died a few days before my arrival; at the end of which, several goats were decapitated as offerings to his manes. All the chief men of the place were assembled there, except Duke Ephraim, who, I have no doubt, secretly laughs at his subjects' folly. There is one thing to be said in his favour, however: the

human sacrifices that were openly performed on
the shore, in the day-time, are now perpetrated
under the veil of night, and are not so numerous
as formerly. I am aware that this is saying little; but it proves, I think, that they are getting
ashamed of their religion. The commander of
one of the Liverpool ships that was lying there
assured me that within his memory these sacrifices
had decreased considerably, but that on the death
of a chief they were still common. The chief,
part of whose obsequies I had witnessed, belonged to a village a short distance up the river; and
his canoe came down after his death, with the
skulls of the men who had paddled her stuck
upon each thwart, to the number of forty. That
such a state of society should exist where British
trade has been carried on to an enormous extent
and for a considerable time, may well excite astonishment in those not aware of the diabolical
influence which the slave-trade exerts both upon
the enslaved and upon the enslaver.

After witnessing the dance, we dined with
Duke Ephraim, who entertains all strangers
every Sunday. The weather was excessively
hot,—and we had an equally hot dinner, consisting of black soup, Calebar chop, palm-oil stew,

and a variety of other African dainties, with
pepper enough in them to have scalded a silver
spoon. After dinner we adjourned to an open
space of ground, to witness a wrestling-match,
between Jacko, a member of the Cameroons Club,
(whom we had brought over for the purpose,) and
the champion of Calebar. The stakes were, two
chests of guns, two barrels of powder, and ten
pieces of cloth, altogether of the value of forty
or fifty pounds ; and as the match had been long
talked of, it excited great interest amongst the
inhabitants. The wrestlers were both powerful
men; but the Calebar hero had evidently been
sleeping upon his laurels, and was too much in
flesh. Jacko was not so tall as his antagonist, but
was, I think, without exception, the broadest man
over the shoulders I ever saw,—his legs and arms
were literally cables of muscles. They were both
oiled all over, and commenced play by approaching each other in a stooping position, scraping
the ground with their hands, to get the sand to
adhere to their oily fingers. The Calebar man
made the first spring; which was skilfully avoided by Jacko, who, as his opponent flew past him,
caught him by one of his ancles, and throwing
him a complete summerset, landed him on his

back with a shock that left him insensible for a few minutes. They were re-oiled, and made play again more cautiously. We now saw an interesting display. Laying their left hands upon each other's shoulders, they alternately caught at the heads and legs of each other. After a severe struggle the victory was again Jacko's, who took his gigantic antagonist under the knee, lifted him fairly in the air, and threw him in the most approved style, his shoulder-blades first touching the ground. Although there were between two and three thousand people present, and their champion had been overthrown, the greatest good humour prevailed, and the two principals walked off the ground, the observed of all observers, in a most amicable way.

On laying the Quorra aground, we were fortunately enabled to replace the copper sheathing which had been rubbed off from her bottom, as in these seas the "*teredo navalis*" abounds in an extraordinary degree. The exposure to which I was subjected while doing this, however, brought on a severe return of intermittent fever, and I was glad to change the close and pestilential atmosphere of Calebar for the clear and salubrious air of Clarence Cove.

A few days after my return, I had an attack of cramp in the stomach, which, if it had not been for the prompt assistance of Colonel Nicolls, who heard me fall out of my bed, would most probably have terminated fatally. This, in addition to my daily fever, confined me for some time, and was the more annoying, as it prevented me from accompanying the colonel to his station upon the hill, which I was very desirous of visiting, as it is considerably above the fever-level.

Mr. Becroft, who had returned to Calebar in the Alfred, a brig that arrived from England belonging to his firm, brought information that a large square-rigged schooner under Spanish colours was lying at the entrance of the river, and that Monsieur Gaspar had the slaves collected for her. A few days afterwards two vessels hove in sight, and proved to be his Majesty's brig Brisk, and a prize, which, upon Lieutenant Thompson landing, we were delighted to hear was the schooner from Calebar, with three hundred and thirty-three slaves on board. The prize did not come in, but stood away to leeward off the island, on her passage to Sierra Leone for adjudication. As usual, she had been stumbled on by mere chance. The Brisk was blockading Bonny, in which river there were eleven slavers. Being out of provisions and water, she was obliged to leave her station, and was running down to Fernando Po for supplies, when she fortunately fell in with the schooner, that had only left Calebar forty-eight hours before, and, afraid of being seen from the settlement of Clarence, had attempted to weather the island. If she had gone the usual route between Cameroons and the island, the probability is, she would have got clear off. *

If Fernando Po had been the seat of the mixed commission court, the slaves would have been landed in twelve hours after their capture, and the prize-crew restored to their own vessel; in place of which, the prize had to make a voyage of two thousand miles, would most probably lose a great number of the slaves, and so weaken the crew of the Brisk, that she would have to proceed to Sierra Leone to get them back, leaving Bonny totally unguarded, instead of being back at her station in twenty-four hours. This may

* Since my arrival in England, I have heard that my acquaintance M. Gaspar got safely off with the next cargo of four hundred and fifty slaves, about a fortnight after this capture.

be stated as the case with respect to five-sixths of the captures that are made in the Bights of Benin and Biafra.

The period of my stay on the island now drew rapidly to a close; and on the 29th of October I took leave of my kind and generous host, Colonel Nicolls, feeling most sincerely that I had incurred, during my residence of two months on the island, a debt of gratitude that I should never be able to repay. To Mr. Becroft and Dr. Butter I am under great obligations, both for their attentions to myself and to my crew, and I take this opportunity of returning them my sincere and heartfelt thanks.

Fernando Po has been alternately represented as an African paradise, and vilified as the most pestilential spot on that unhealthy coast. I found it neither the one nor the other. That it is much healther than the main land is obvious; for one is a swamp—the other a dry and elevated island, exposed on all sides to the sea-breeze, and free from that excessive humidity which so fatally distinguishes the neighbouring coast. During my stay there, the crews both of the brig and the steamer recovered rapidly; and individually I should have left the island in comparative health, had I not exposed myself at Calebar and relapsed in consequence. I certainly lost one man while lying there: he went on shore, got drunk, and slept upon the grass, exposed to heavy rain,—which would kill most men in any country.

Fernando Po has acquired the reputation of being unhealthy and pestilential from two causes: — one, that Europeans were employed as artificers in the erection and construction of the buildings, instead of coloured people. If it be difficult in England to keep mechanics steady, it is impossible to do so in a tropical climate; and if to this be added the exposure while clearing the ground of the jungle, it will excite no surprise that the residents suffered so much from sickness, and lost so many upon their first landing. The other cause of complaint is, that a great part of the mortality that takes place at Fernando Po is from diseases contracted on the main land, which in common fairness should not be charged upon the island. Several Europeans died in the hospital during my residence there; but they had been landed from ships, in the last stages of disease. Leaving the mortality created by these two causes, — one of which (the clearing, &c.) will not occur again, and the other, which, being a foreign one, should not be taken into the calculation, the mortality of Fernando Po will be found much less than that of other settlements upon that coast, if it be not upon a par with the most healthy of our West Indian colonies.

Fernando Po has one advantage over any other situation upon the coast: its elevation is so great, that any climate may be obtained there, and an hospital might be established above the fever-level. In fact, Colonel Nicolls had cut a road up the mountain, and built a small house above the fever-range; and I had the pleasure of seeing at Calebar, Mr. Ballard, a gentleman who was carried up in a hammock, in the last stage of fever, and had recovered immediately. The fact that marsh-fever does not exist above a certain level (I believe three thousand feet above the sea) is well known; and as all African fevers are varieties produced by marsh miasma, Fernando Po may be said to have been placed by Nature in its present position as the natural hospital for the low and marshy shores of the main land. The native population of Fernando Po may be about five thousand, divided into tribes, that were formerly constantly at war with each other; but through the exertions of Colonel Nicolls, and the influence he possessed, combined with the confidence they repose in the justice of his decisions, referring every dispute to him; they are now living in peace and amity.

They are a well made athletic people, rather under the middle size, with a much more intelligent countenance than their neighbours on the opposite coast. It has been said that they have a cross of Portuguese blood in them, from the crew of a Portuguese vessel that was lost on the island in the early ages of discovery. Of this I think there is no proof, unless their abominably dirty habits may be adduced as establishing the relationship; the Portuguese being the dirtiest of European nations, and the Boobies (the name of the inhabitants of Fernando Po) the dirtiest of the blacks.

They cultivate yams in large quantities, and are great fishermen; they are exceedingly fond of the entrails of fowls and sheep, which after slapping upon their shoulder a few times, they draw across the fire and eat with great relish. Small deer abound in this island; and Colonel Nicolls was regularly supplied with them by the natives, who take them in snares, and claim in return a little powder and shot to shoot monkeys, which form their favourite food: deer they do not eat, from some prejudice or superstition, much the same perhaps as that which makes us prefer the deer to the monkey.

I do not think that slavery is known amongst them; they certainly do not trade in human flesh, and have several times had bloody engagements with slavers who have landed and attempted to surprise and carry them off. This may explain their free and independent bearing, contrasted with the abject demeanour of the Calebar and Cameroons people, to both of which places I took my friend, the chief Cut-throat, and several of his tribe, and was much struck with the comparison.

The improvement which has taken place in the island, is not visible in the outward appearance of the natives, as they still go perfectly naked, and plaster themselves over with oil and clay. But as they are beginning to collect palm-oil and bring it to market, there is little doubt that the comforts and conveniences of civilised life which they receive in return for produce will silently and gradually improve their condition.

Fernando Po is of volcanic formation, in common with the other islands in the Bight of Biafra, and most probably forms part of the great volcanic chain which appears to run from the parallel of the Cape from south-south-west to north-north-east; St. Helena being the centre, and Cameroons and Tristan da Cunha the extremities.*

The incidents of a winter passage home with a sickly crew can afford little to interest, and I shall merely state, that after leaving Fernando Po I encountered adverse winds and currents until I reached Annabona, off which I lay-to for two hours. On the 28th of November, having been thirty days in beating down the African coast, we found ourselves in the south-east trade-wind; on the 30th of December we made Tuskar lighthouse in the St. George's Channel, and on New-year's Day arrived in Liverpool completly worn out, having encountered nothing but gales of wind from the time we left the north-east trade.

On the 4th of December we committed to the deep the body of William Davies, one of the men belonging to the Quorra, who was seized with madness three days after leaving Fernando Po; and on Christmas Day, one of our best hands was disabled by a fall from the main-topsail yard-arm, while taking in the third reef.

* All the Atlantic islands are said to be of volcanic origin.

He had a most miraculous escape both from drowning and a more cruel death. The vessel was scudding under a double-reefed topsail and foresail, and going about nine and a half knots, with a heavy sea running, when I sent the men up to take in the third reef of the main-topsail. He was out on the yard-arm next the weather earing, when the sail got loose and flung him upon deck between a spar and a gun, where he lay insensible, and I thought dead, for some time. On taking him into the cabin, I stripped and bled him, and, to my surprise, found he had not a bone broken. After rubbing him over with brandy, I gave him a dose of laudanum; and by the time he got to Liverpool, he was able to walk. He owed his life principally to the vessel rolling to port at the time he fell;—his extreme fatness was also much in his favour.

Belfringe, one of the Quorra's crew, died a few days after landing in Liverpool, from exhaustion and general debility.

MR. OLDFIELD'S JOURNAL.

CHAPTER I.

*Commencement of Sickness in the Vessels of the Expedition.—
Symptoms of the Fever.—The Boatswain and Commander
of the Quorra die.—Remain in the Columbine.—Departure
of the Steamers up the River.—The Brig Susan.—The Bar
of the Nun.—Landers' Journals.—Loss of the Susan.—
Town of Cassa; character of the Natives.—News of the
Steamers.— Set out to visit Brass Town.—King Boy's
Canoe and its contents.—Fish Town.—Superstitions and
Ceremonies. — The King's Daughters. — Preparations for
Arrival at Brass.— Landing.— The King's Residence.—
Observations on Brass Town.—Funeral Ceremony.—Des-
potism of the African Chiefs. — Return on board the Colum-
bine.*

HAVING accompanied the expedition, the proceedings of which have been related in Mr.
Laird's journal, to the period of the departure
of the two steam-vessels up the Nun, the few
remarks I might have to make on our passage
out have been principally anticipated by that

gentleman. I may briefly state, that my appointment as junior surgeón to the party took place at Liverpool, and that I had scarcely arrived on the coast in the Alburkah when my professional services were put in requisition by the effects of the climate.

After leaving York in the Alburkah, the first person attacked was Mr. M'Kenzie, the boatswain: Captain Hill, and Smith, a seaman, were also attacked with the same symptoms of epidemic fever. The boatswain appeared to labour under a presentiment that he should not recover, and giving himself up to melancholy, desired to be left at York, as he preferred dying on shore to doing so at sea. Notwithstanding this unfavourable condition of his mind, he recovered like the rest, but was again seized with fever, and died within a week of his first attack. *

* The symptoms of the fever did not vary much in these cases from those which occurred on board the Quorra at the same time, as Dr. Briggs and I found on comparing our notes. The patients first complained of giddiness, slight pain across the forehead and at the back of the orbits ; pulse quick, tongue white and red at the tip, obscure vision, great prostration of strength, sometimes vomiting a dark-coloured matter, and great thirst. After the first few hours, the patient appeared anxious to avoid all communication with any one, and remained in his hammock, or concealed himself in

The day after our arrival at Cape Coast, I found that Dr. Briggs had three officers and several men sick in the Quorra, and had landed a man named Morgan in the care of the surgeon of the fort, who afterwards died. The Columbine had two sick on board at the same time, and in the Alburkah we had the chief mate and four men sick.

In crossing the bar of the Nun river, the Alburkah was drifted by the current over to the eastern breakers, and was in great danger of being lost. Being much in want of fuel, we were compelled to burn every disposable spar we could find : even the jib-boom of the vessel, and boxes of all kinds were condemned to the flames in order to obtain sufficient steam to enter the river. By this means we succeeded in doing so,

some obscure part of the vessel. When asked to state how he felt, he would reply that he was quite well, and denied that he suffered any pain. About the third day he became delirious ; and from the fourth to the ninth day, death put an end to his sufferings. The three patients alluded to were cupped and blistered, their heads were shaved, and cold lotions were applied, purgatives administered, and afterwards five or ten grains of calomel given every four or six hours, until ptyalism was slightly produced. Smith being a robust man, was bled to thirty ounces, and lived several months afterwards. He died up the river, of dysentery.

and anchored near the Quorra and Columbine, at the expiration of eighty-five days from leaving England.

The night before we entered the river, Mr. M'Kenzie, the boatswain, died. He had been seized a second time with fever. The day before he died, he came on deck and appeared better. About seven in the evening, I happened to go into the forecastle, and found him lying on some wood, quite dead. Poor fellow! he was committed to the deep the same night with the proper ceremony. The other invalids became better, although his death tended greatly to depress their spirits. The next morning a boat from the Quorra visited us; and I found that while the sad ceremony was going forward, the same was performing on board that vessel over the remains of Captain Harries. The command of the Quorra had devolved in consequence on Mr. M'Gregor Laird, who held a considerable share in the expedition.

On the 21st of October, I was engaged with Dr. Briggs in selecting and packing medicines for the two steam-vessels, to take with them up the river. The opinion was that they would be down again in two months; but Dr. Briggs very

properly determined on taking enough for five,
in addition to what each vessel had in her medi-
cine-chest. From the day of our arrival in the
river, to the departure of the vessels, every one
was employed in transhipping the goods on board
the Quorra, and supplying the vessels with coals
from the Columbine. It had been arranged before our departure from England, that the junior
medical officer was to remain on board the Columbine, at the mouth of the river, until the return of the steamers from the interior, which
was then expected to be in about two months;
and accordingly I took up my quarters on board
the brig, congratulating myself at the same time
that I had enjoyed the good fortune of being
one who had crossed the Atlantic in the first
iron steam-vessel that had ever been sent to
sea.

On the evening of the 25th of October, the
Quorra and Alburkah being quite ready for ascending the river, I took my leave of Mr. Lander, and they departed under the hearty cheers
of the crew of the Columbine, which those in the
vessels returned with the usual spirit of British
seamen.

The excitement of the enterprise now gradually subsided in the minds of those who were left on board the brig. The presence of the vessels which had just departed, had served to keep not only our heads, but our hands employed; and as they were now gone, we had nothing to do but to lie still at our anchorage, and dwell on the expectation of their return at the end of that time which we had allowed them for their voyage.

The first object which naturally attracted my notice was the brig Susan, alluded to by Mr. Laird, that was lying in the river, affording a melancholy proof of the fatal effects of the climate of this coast. She had been there nearly nine months, and had received her cargo of palm-oil; but in consequence of the deaths of the crew, consisting of nine persons, the captain, who was also reduced to a skeleton, was unable to move the vessel. He had purchased his palm-oil at the rate of seven to nine pounds per ton,—a reasonable price for the coast. I believe that oil may be had much cheaper near this river than any other; but from the imperfect knowledge which our vessels have of the passage over the bar, it is not much frequented by them. I have frequently sounded on the bar, and never found less than a quarter less three fathoms; and at dead low water of spring tides, have found four fathoms.

The reader is probably aware that Messrs. Lander, in their celebrated expedition, lost their journals as they were coming down the river, and afterwards obtained a passage on board the brig Martha, without having it in their power to remunerate King Boy for ransoming them at Eboe, owing to the strange conduct of the captain of the Thomas. King Boy had found these journals, and sold them to Captain Townson of the Susan, now in the river, for eight puncheons of oil, or goods to purchase that quantity. On our arrival King Boy stated to us that he had given the books (journals) to the cappy (captain), who had only given him in return for them the price of three puncheons. A great deal of dissatisfaction manifested itself in King Boy, and he complained lustily of the very little he had received for them. Captain Townson still persisted that he had given Boy the price of eight puncheons, and Mr. Lander presented him with a bill on Government for the amount. However, it was not forgotten by Boy; for one afternoon, when Townson was in the cabin of the Alburkah,

King Boy came on board. He seated himself next to me, Captain Townson being on my right, and began to complain of having received only three puncheons. Captain Townson said he had given bars or goods to the amount of eight puncheons. King Boy said, " No, no, cappy; you lie, lie, lie !" Townson, who was but a little man, forgot himself so far as to reach over me and strike his sable majesty a violent blow on the face, so as to cut the inner side of his lip; and a scene ensued which reflected no honour on the character of the commander of a British merchant-ship. Boy was not a little sulky, and uncommunicative for the remainder of his stay; but he seems to be an avaricious man, grasping at all he can get, and yet without appearing satisfied. Mr. Lander had made him several very valuable presents— fifteen or sixteen guns, two barrels of gunpowder, fifteen soldiers' canteens, knives, spoons, and soldiers' coats, with various other articles, with which he was not contented.

Captain Townson was in great want of provisions and men. Mr. Lander kindly ordered two buckets of cocoa and coffee to be given him; and two men who appeared unwilling to proceed up the river, left the Quorra to join his vessel. I was also desired to supply him with medicine and advice : I accordingly visited him and his sick two or three times a day during his stay in the river.

About the 6th or 7th of November, the Susan having received assistance from the Columbine, got under weigh and crossed the bar ; in doing which she struck and started her stern post, but suffered no other damage. She was afterwards deserted off Prince's Island, and the wreck floated into Bimbia and Cameroons, the natives reaping a rich harvest from the palm-oil she had on board. It was subsequently reported that the brig had met with foul play, and that she had been scuttled. She was evidently saved by us but to be lost ; for in all probability, had we not arrived when we did, she would have remained in the river until she decayed. But the character of Captain Townson was not the most humane. A man named Fanby had joined the brig Susan from the Quorra steamer, who, after being treated by him in a most brutal manner, was turned adrift at Cassa or Pilot Town for some misconduct. As soon as we heard of it, he was taken on board the Columbine in a deplorable condition, having been ashore nearly a fort-

night. The natives among whom the captain
had landed him at Cassa were little better than
savages, and in consequence he was almost starv-
ed to death, being obliged to subsist on bananas and dead rats. I afterwards had many
opportunities of witnessing the rat-feasts of
these natives,—and disgusting sights they are.
However, Fanby never recovered his accustomed
flow of good spirits; something appeared to be
preying on his mind from the time he was taken
on board the Columbine. He behaved himself in
a very orderly manner to the time of his death,
which took place in a few weeks. The day before
he died, he complained of being very unwell,
but could not describe any particular pain: he
was on duty on deck in the evening when I went
into the forecastle, and he conversed with me
very rationally. At four the following morning
he went on deck, and at five was a corpse. Poor
fellow! his remains were interred on a beautiful
island at the mouth of the river, surrounded by
the sea, but which has since been completely
washed away.

The Columbine was moored at the mouth of
the river, open to the sea-breeze, with one hundred and thirty fathoms cable.

Dido, the pilot of Cassa, frequently visited us, bringing yams, cocoa-nuts, and fowls for sale. The town of Cassa, to which he belongs, consists of about forty huts, irregularly built among plantain and banana trees. There are also a great many cocoa-nut trees, the property of the natives. The method adopted by these people in ascending these lofty trees, is as follows :—A piece of the bark of a tree is twisted into a rope, and two pieces of it are used; one is secured round the body of the tree and right shoulder of the native, the other round the left thigh and footing; he then raises his leg and foot, supporting the upper part of the body, and gradually slips up.

The natives of Cassa are a most indolent people. Some few of the women paddle their own canoes, and are very expert at spearing fish. They also subsist on dogs, which they fatten for the purpose, and consider monkeys a delicacy; and although they can boast of a few fine Spanish sheep, these are strictly kept as fetish, or sacred. They frequently brought us fish for sale; but goats and sheep they could only obtain from the Eboe people. Polygamy is common all over Western Africa, each man calculating on his women as property according to the trade they make.

Dido has four wives, three of whom live with him at Cassa, and one at Brass Town. He is a slave of King Jacket's at Brass, and resides here for the purpose of conducting vessels into the river. He is a very shrewd, intelligent man, and speaks good English. In all our dealings we found him strictly honest. In his visits to us on board the Columbine, this pilot generally brought us the gossip of the place, which he dealt out with an air of great importance. Among other stories, he told us of a report that the vessels had been attacked in the little Eboe country, and some white men killed. What degree of confidence to put in this statement, I did not know; but it appeared to me not unlikely to be true. We now became every day more anxious for the return of King Boy from the steamers with despatches. The last few mornings had been excessively hot, and we had had several tornadoes from the eastward, the good effects of which in clearing and purifying the air were evident to us.

On the morning of the 13th of November, we were agreeably surprised by the appearance of King Boy, who brought us accounts from the steamers. I received a letter from Captain Hill, which partly confirmed the report of the pilot with reference to the affair which has been related by Mr. Laird.

Being very desirous of visiting Brass Town with the captain of the Columbine, after preparing some blanket-dresses, and providing ourselves with everything we had that was necessary, we rowed up to Barracoon point. Here we found King Boy's state canoe lying, and also four of his wives, whose names were Bella, Awanhee, Nain, and Hoowarrh. There were also seven slaves, a small bullock, and a goat and pig, at this establishment.—It would be almost impossible to describe the various articles with which this king's state canoe was loaded, most of which were presents from Mr. Lander. Including thirty pullaboys, and the slaves, wives, and other attendants, there were upwards of fifty individuals in the canoe, which was in consequence exceedingly deep, and her gunwale nearly even with the water's edge. In consequence of this, two slaves were continually baling out the water as fast as it found its way in. There did not appear to be much regard to order or good feeling among those in the canoe, for I was concerned to see one of the female slaves crushed for want of room: in her arms she held an infant about five days old, and both mother and child appeared to have had their heads shaved that afternoon. Among the rest were six men, natives of the Eboe and Ibbodoh country. Each country has its peculiar distinguishing mark which is adopted by its people, and the latter had a longitudinal incision from the external angle of the eye to the mouth. I learnt from King Boy that he pays two muskets, two jackets, and a few yards of printed cotton, for each of his slaves, and that two slaves are equal in value to a puncheon of oil in the Eboe country.

As the object of our visit was to accompany King Boy to Brass, we embarked in his canoe, and were placed in the centre of the fore part of it.* Immediately below us were barrels of powder, cutlasses, guns, cans, small swivels, poultry, yams, a wild pig, and a small bullock about the size of an English calf, all the property of King Boy; and surely never had he reaped such a golden harvest as he had now done from the two steamers, and was carrying off to his abode.

* In the stern of the canoe a large fire was kept burning on some wet sand.

The pullaboys were seated on each side of us; a mat was spread for us to lie upon, and a pole ran fore and aft in the centre of the canoe, over which was placed another mat to prevent the dew or rain from wetting us. Having on my blanket-dress, and being wrapped in my cloak, I lay down in the canoe after taking the precaution to place a union jack over our heads in the form of a curtain as a protection from the mosquitoes, and we set out on our voyage. The pullaboys appeared to be greatly fatigued, having been paddling four days and nights on their passage from Eboe. I found from Boy that a Brass canoe takes seven or eight days to reach Eboe from Brass, and four to return. One of the crew, who I found was dignified with the title of King Boy's mate, amused us much by singing a country song. As he chanted each verse or part, the pullaboys would repeat the same words twice over, and then, after flourishing their paddles over their head, would dash them into the water, and applying all their strength, propel the canoe with great velocity. We had besides a current in our favour of about three and a half knots per hour; so that we got on tolerably well.

Previously to our leaving the Nun, we were
given to understand that our journey would be
ended by sunrise on the morning after we had
set out; and as King Boy promised us plenty of
chop to eat, we had merely taken with us a few
little biscuits. About eleven o'clock we stopped
at a town on the left bank, consisting of a few
huts, named Fish Town. This town is inhabited
chiefly by fishermen, who dry large quantities of
fish. The method of drying them is by placing
them about ten feet above a wood-fire in the
smoke for a few hours. A great number of
oyster-shells were lying about the bank, which
the natives calcine and convert into a good chalk.
The pullaboys remained here about two hours to
rest. Having had nothing to eat since we started, I found the cravings of nature so powerful,
that I gladly partook of what I could not have
been induced to do at another time. This was
a cold piece of boiled yam saturated with Mala-
getta pepper, which tasted as unpleasant as cold
potatoe dipped in vitriolic acid.

About one o'clock we proceeded on our
journey. The night was fine, and excepting the
occasional strange cries of the pullaboys, and the
noise of their paddles entering the water, all was

silence around us.   The sides of the river which
we passed were always low and swampy: no
banks or towns were to be seen.

About five o'clock I was aroused from a pleasant sleep by the firing of guns.   The least noise
or movement generally disturbed me; and as I
did not feel quite confident of Boy's sincerity, I
inquired the cause of the firing at such an unusual hour, and was told by the black mate, a
good-hearted fellow, it was firing for Ju-ju.   It
seems that the Brass traders never pass certain
places without firing, and repeating certain
words as charms.   When they take rum, they
pour a little on the deck or in the river for the
Ju-ju, and they would consider their lives very
insecure if this ceremony was not observed.
Whether the strange scene about me, or the
effluvia proceeding from so many negroes in a
state of nudity perspiring from exertion, or from
being so suddenly aroused, prevented my falling
asleep again, I know not, but I could no longer
close my eyes.   At one time we must have
passed very near the sea, as I could hear distinctly the noise of the surf on the beach.   We
crossed a great many branches of rivers, all of
fresh water, intersecting larger ones, and fall-

ing into the sea from their parent stream, the Niger.

In the course of the morning, we saw, at a short distance from us, two small canoes with females paddling them. They came alongside, and I found, on inquiry, that these damsels were the daughters of King Jacket, of Brass. King Boy poured out two glasses of rum for each, which these princesses drank off without moving a muscle of their black faces ! and with the same ease that one of our dairymaids would have drunk a basin of new milk. The youngest lady appeared to be about thirteen, and the elder fourteen years old, and both were fine, good-looking girls. Their skins were of a lighter colour than any I had seen; but their sense of modesty did not appear to be shocked in the least by their naked condition, for they knelt down in the bottom of their canoes without the least concern at our presence, although after drinking the rum they glanced very earnestly at us as if we were somebody unexpected.

At eleven o'clock, being near to Brass, preparations were made to fire the two swivels given by Mr. Lander, for the little Ju-ju, as

King Boy called it. Accordingly the swivels were fired; and on my inquiring where the Ju-ju was, I was shown a fine majestic tree in which, I was gravely told, he resided.

We were now about two hours' journey from Brass Town. When at a short distance from a creek leading direct to the town, the canoe was stopped, and every one in her began to make preparations for their arrival. King Boy, with his boys, jumped overboard for a swim, and was speedily followed by three of his wives. While they were bathing, the fourth wife was employed in selecting the clothes from the royal wardrobe.

As this was the first time King Boy had been at Brass since leaving Mr. Lander, it was the first opportunity he had of showing off his splendid presents. Great fuss was made to do justice to himself: he said to me, " King Boy leave Brass all same one little boy, no guns, no money; but now me big man, big man too much, and plenty ting." The once splendid dress, sent out as a present from Mr. William Laird, of Liverpool, was now taken out of a large calabash, not a little the worse for being kept there. It consisted of a scarlet coat, a Highland kilt and goat-skin purse, plaid stockings and slippers, a tartan plaid sash, and lancer's cap.

After he had been dried by his attendants, he proceeded to the more important duties of dressing himself; and as his style of dressing was ludicrous in the extreme, I must attempt a description of it. A blue handkerchief was first tied round the loins, covering the right leg : over this was placed a striped shirt, then a pair of Turkish trousers, and then the plaid stockings and slippers (the Turkish trousers completely concealing them). He next put on his scarlet coat, the waistband of the trousers covering the lower part of it; over the trousers the goat-skin, while the plaid scarf was thrown over the left shoulder. On his head he placed a red worsted cap, on which the lancer's cap was fixed. Through a serjeant's sash, and a goat-skin round his middle, he thrust an old cutlass and one pistol. And this grotesque dress, contrasted with his sable features, had the effect of making him appear, to the eye of a European, the most ridiculous of objects—a curious compound of vanity and absurdity.

While this was going forward, the mate kept up a constant fire with the swivels, and we ap-

proached Brass Town, which is situate on a morass, with scarcely one yard elevation above the
bed of the river, while the natives assembled to
witness the grand sight presented by our arrival.

King Boy now began to consider himself a
king indeed, and applying a speaking-trumpet
about a yard and a half long to his lips, as he
stood in a conspicuous part of the canoe, bawled
out with stentorian lungs, " Devil-ship (meaning the steamers) no will! King Boy no will!
Devil-ship no will! King Boy no will! King
Boy all same King Obie! King Boy big, all same
King Obie! Palaver set — palaver set!" with
a great deal more about himself, at which I was
highly amused.

The first place we stopped at was a few
hundred yards below King Boy's residence, at
the Ju-ju man's, or priest's, hut. This person
wore a small piece of white calico round his
loins. He waded up to his knees in mud,
and received the presents sent him from Mr.
Lander. I do not know whether all were given
him or not, but Boy gravely delivered over to
him a red coat, a bottle of rum, and a jug made
after the figure of a Greenwich pensioner.—On
the bank, near the canoe, sat a shrivelled old

woman, the very picture of misery, with her breasts dried and drooping in a most disgusting manner.   But this old hag drank six and a half glasses of rum that were given to her undiluted! and as she swallowed each one, began to caper and dance in the fandango style, throwing herself into the most indecent postures and contortions possible.   I felt curious to know who she was, and I found that she was King Boy's mother.   We left the place in a few minutes, and proceeded to the residence of King Boy, where we landed, and entered a house open at its sides and roofed with dry grass and bamboo (*Arundo bambos*).

This building we found was used by King Boy's boys and slaves.   Opposite to this, and on the bank of the river, were his houses, three in number.   The centre one was the apartment in which King Boy slept, and was elevated above the others about three feet: one dark passage communicated from one end to the other.   They were built of bamboo and plastered with mud, being one story high.   Such was the palace of King Boy!   The interior was frightful to behold: it was unpaved, and the floor of mud quite wet from the encroachments of the river, which is sometimes ankle-deep in the first house of the three; added to which, it was perforated with numerous rat-holes. At one end of the apartment, near the river, was an elevated mud platform in the shape of a triangle, just as damp as the floor. As stools or chairs formed no part of the furniture of this palace in the swamp, we procured as substitutes a quantity of mats to lie down on.

As soon as the numerous articles belonging to King Boy were landed from the canoe, he invited us to go with him and partake of some goat's flesh; and we therefore accompanied him through several narrow, filthy streets, to the house occupied by his favourite wife, where, for the sake of security, he invariably has his meals. After some little delay we were accommodated with a knife, when some boiled goat's flesh with the hair upon it, and highly seasoned with Guinea pepper, was placed before us. The accompaniment of the hair I did not much mind, but the pepper I could not endure. When King Boy saw I could not eat it, hungry as I felt, he very considerately ordered some more to be cooked without pepper, and to be better cleaned. But I was again doomed to disappointment; for my head ached with such violence, that I could not eat or move a muscle of my face. I was very apprehensive of being laid up on the spot with fever; and certainly, neither before nor since, have I ever experienced such an acute, dreadful pain : tic douloureux was nothing in comparison to it.

My chief resource was smoking; and in the evening Captain Mitchell and I went to King Boy's houses on the bank of the river, to pass the night. Having spread my cloak, and with a sword and pistol by my side in case of treachery, I laid myself down. A small, hollow, wooden dish had been brought to us, filled with palm-oil, and a piece of the bark of a tree was placed in it for a wick. Having requested this primitive lamp to be taken away, it was no sooner done and we were left in darkness, than a noise commenced as if people were attempting to get in through the roof of the house. I instantly rose and called out, when the noise ceased, but soon recommenced; and, hearing the squeaking of rats, we concluded that they had created the disturbance. They were large water-rats, whose nests were in the roof of the house,

and the floor was quite undermined by them.
From the incessant noise they made, their number must have been immense.

In the course of the afternoon, one of the
pullaboys had stolen a pocket-handkerchief from
me, and I mentioned the circumstance to Bella,
the king's eldest wife, through an interpreter.
She seemed to express much concern about it,
and examined them all before her, and held a
long palaver, the substance of which was, that I
had lost a handkerchief, and that she insisted on
its being delivered up on pain of her displeasure.
After being out a few minutes, I returned, and
found my handkerchief lying upon my cloak,
evidently placed there by the guilty party.

Brass Town, I have before observed, is situated in a morass; the streets are very narrow,
and abominably filthy;—neither before nor since
have I ever met with a more disgusting place.
The tide comes up to the town, and at low water
the filth of the natives is indescribable.   I walked
a short distance into the bush; the soil was alluvial.   The natives subsist chiefly on fish, plantain (*musa sapientum*), and bananas (*musa para-
disalia*), which are brought from the little Eboe

country, very few of these ever growing at Brass. Tigers and leopards are said occasionally to come into the town and carry off children.

Among the sights of the place, were two natives with their faces whitened, and two white feathers stuck at each side of their heads. These fellows go about dancing in a most singular manner, and imitating at the same time the roaring of wild beasts. I was told these men were leopard-hunters, and were painted white, with the feathers for horns, to represent the devil : for, as we believe Satan to be black, the Africans, on the contrary, represent him as being white.

While we were here, I had an opportunity of witnessing a native funeral. An elderly man had died a few days before, and was laid out on the floor of his hut. He had left four wives, who were crying and making a great noise over the corpse, as usual on these occasions, and apparently in great distress. The corpse was enveloped in upwards of one hundred fathoms of calico—two hundred yards ! It had been rolled round the body very tightly, from the neck to the toes ; the head was uncovered, and the whole corpse presented the appearance of a black man's head stuck upright in a bale of cotton. At the

head of the corpse was a bottle of rum, a calabash of palm-wine, two live fowls, and some bananas,—all which were to be interred with it,
to keep it from starving in another world! On
one side of the hut, which was not capable of
containing more than a dozen people, was a
grave, about two feet deep and four wide, with
water at the bottom of it: being so near the
bank of the river, it was impossible to dig deeper
on account of the appearance of water. Several
singing-men and women were outside of the hut,
following their avocation, beating a broken piece
of canoe with sticks, and making the most stun-
ning noises; on the other were some traders,
just arrived from Bonny to buy palm-oil, and who
were serving out rum with an unsparing hand;
while at the back part of the house people were
continually firing muskets, and two old honey-
combed four-pounders, in a barracoon or slave-
house, much to my terror lest they should burst.
I understood these ceremonies would be continued several days; and on the last day one
of his four wives, would be thrown into the
river as a sacrifice, instead of being buried
alive with the corpse, as is the custom at
Calebar.

The day after our arrival, King Boy was endeavouring by every method in his power to obtain more presents from us. As we had had quite enough of Brass Town and its inhabitants, we felt anxious to depart; but not having a canoe or boat of our own, we were obliged to depend on the generosity of King Boy to supply us when he thought proper; and, partly by menaces and entreaties, we obtained the promise of a canoe. But it is not unlikely that this promise was partly hastened by a threat, that unless he let us have a canoe, I would apply to King Jacket, of whom Boy was very much afraid. Jacket possesses the most absolute authority, and occupies one bank of the river; the opposite being occupied by King Fourday and his people, as well as by Kings Boy and Gun, sons of King Fourday. Gun, the little military king, as the Messrs. Lander have styled him in their work, is a pleasant, good-humoured little man, and perhaps the most honest in the town of Brass. King Fourday is a very old man, and very much addicted to rum. When this sable potentate paid us a visit on board the Alburkah, three weeks before, he drank so much that we were obliged to lift him into his canoe.

Louis, the pilot who accompanied us in Boy's canoe, is a shrewd intelligent negro, and, from his intercourse with Spanish vessels, speaks the Spanish language remarkably well. In the course of the day, while we were at Brass,* he came to me and requested me to go and look at his house, which to gratify him I did. He produced some excellent tombo, or palm-wine; and afterwards, pointing to a shrub, and a jar with water, and a small piece of yam in it, said it was his Ju-ju.

The following anecdote will serve to illustrate the absolute power possessed by these African kings:—King Gun had accidentally received a wound in the arm from one of two boys who were playing with cutlasses. This boy threw his cutlass at the other: Gun happened to be passing the corner of the street at the same time, and received the point of the cutlass in his arm. The boy was immediately

---

* This man, Louis, was afterwards tied to a piece of wood and thrown into the river to the sharks, for his intimacy with one of King Fourday's wives. The method of punishment for crimes of this nature consists in the parties being drowned, or obliged to take poison. In this instance, the latter was preferred by the unfortunate woman, while the pilot chose the former mode of death.

seized, and begged and entreated mercy to no purpose. A council was held, and he was condemned to die : his hands were tied behind him, and his throat being deliberately cut, he was left to bleed to death!

Captain Mitchell and I having seen quite sufficient of Brass, were very glad to embark in Boy's canoe at night, and leave the place. On the next morning we safely reached the Columbine.

CHAPTER II.

ABOUT a week after my return from Brass, I was seized with a fever; and finding it very strong upon me, I abstracted twenty ounces of blood from my left arm, which gave me consider- able relief. On the third day I suffered a little more pain in my head, and felt very unwell: the symptoms were the same as already described. I gave orders for my head to be shaved and blistered, and also the nape of my neck, and commenced taking twelve grains of calomel in three doses every six hours. In about seventy-six hours after the attack, I became delirious, and was informed that I continued so for nearly a fortnight, attempting to throw myself overboard; a feat which I once had nearly performed, and was only prevented effecting my design by the watch on deck. I continued in a very precarious state for twenty-nine days, during which time I took nothing but water, with a little biscuit soaked in it.

In six weeks I became convalescent, and was much indebted to Mr. Robb, the chief mate, who paid me great attention during my illness. To such a low state had I been reduced, that it was nearly three months before I could walk; but it pleased the Almighty to preserve me, and after this severe fit of sickness, during which I had been several times at death's door, when I recovered, I enjoyed better health than I ever enjoyed in England. I was troubled with attacks of intermittent fever; and before I reached England, upwards of two years afterwards, I had had more than one hundred attacks of it: but this was nothing compared with jungle fever. As I was recovering, one of the men of the Columbine was taken ill; and although I was obliged to be supported from being so weak, I bled him, and before the operation was finished I fainted three times. After the decease of James Fanby, who died in January, this was the first death on board the brig, and it occasioned great despondency throughout the crew.

In a month after the decease of Fanby, the sail-maker, Mr. Jeffrey, died. He was an old pensioner, had been at the battles of Trafalgar and Copenhagen, and served under Lord Nelson. He was fifty-eight years of age, and had often expressed a wish to live until sixty.

In the beginning of January, we had long been anxiously expecting the steamers down the river, but could learn no intelligence respecting them. They had now been absent three months, and our situation was every day becoming more painful, being exposed day after day to the same degree of monotony from morning till night. Our provisions were also getting short; the water was exceedingly bad, and quite brackish.

On the west side of the river, near where we lay, is a barracoon or slave-house, where we found several dozen human skulls lining the bank. I soon discovered that these were the remains of slaves who had died here, this being the place

from which the Spanish vessels take in their cargoes of human traffic.   Barracoon Point forms a
kind of bay, and the tall trees lining the bank
effectually conceal any vessel from the view of a
man-of-war off this river; and unless the boats enter it and go up to the point, the chances are that
the slave-dealers are never seen.   There is also
a narrow creek above this which is highly favourable for concealment.   I have been informed that
two boats from an English man-of-war came into
the river and landed at Cassa, or Pilot Town, on
the east side, without any suspicion of a slaver
being concealed in the above creek : they shortly
went away again, and on the following day the
slaver took her slaves on board and made her
escape.   Opposite the house, within a few yards
of it, a well has been sunk, about three feet deep,
the water of which produces diarrhœa and dysentery with slight attacks of fever.   I recommended
all the water to be boiled before use; and after
that was done, we enjoyed better health.

About this time I received a message from
King Jacket, of Brass, stating that our being anchored near the mouth of the river prevented
slave-vessels from entering, at which he was not
pleased.   But the natives were very anxious to

make trade with us, as they termed it; and King
Boy, who came on board occasionally, was, as
usual, always begging for everything he saw. He
told us frequently, in an exulting manner, that he
could supply a ship with fifty puncheons of oil in
a month, or one moon. It will be recollected the
Susan filled in two months, or nearly so, when
she lost her men; but oil is now very plentiful, as
great quantities of it are brought from the Eboe
country by the Brass traders to Brass Town,
where it is purchased, as well as slaves, by the
Bonny traders. The oil is obtained in the Eboe
country for four or five pounds per ton. The
best goods for this trade are muskets, powder,
red beads, white baft, common scarlet cloth, blue
beads, Bandanas, romals, coarse stuff hats, pipes,
tobacco in leaf, and looking-glasses. A puncheon
of oil is termed so many bars, varying according
to the state of the market: a gun is six bars;
a head of tobacco, two bars; and so on in proportion. Cowries are taken at Eboe, and all up
the country. The Brass traders, like most other
black traders, are not to be trusted; they require to be closely watched, and the commander
of a vessel should not allow them to gain the least
ascendency over him. Having once obtained

a great many goods, they sell the oil which is his to the Bonny traders at Brass, receiving their goods in return. It is well known that Bonny a few years back could scarcely fill one vessel with oil; but at present there are from seven to nine annually in the river; while the Nun, from its proximity to the Eboe country, is likely to produce larger quantities of oil. King Gun, King Jacket, and one or two others at Brass, are the best and most honest traders. King Boy ought not to be trusted too far, for I believe him to be a wicked, designing, and treacherous villain.

February came, and we had no accounts of the steamers: our provisions had become very short, when the William Rathbone was passing the bar on her way to Bonny. Our long-boat being out at the time, we boarded her, and Captain Moore very kindly supplied us with a few acceptable articles.

About the 5th of March, Captain Mitchell, who incautiously exposed himself to a violent tornado and afterwards neglected putting on dry clothes, was taken violently ill. During the last month I had caused a house to be built on Columbine Island, at the mouth of the river, open on all sides to the sea-breeze, for two or

three invalids, where the steward and a boy of
the brig had speedily recovered.   When Captain
Mitchell was taken ill, he felt a great wish to be
landed there ; and a temporary house was built of
bamboo, and covered over with mats, and he was
conveyed ashore.   I remained with him three
days and nights, during which time it was an
utter impossibility for me to obtain any rest
from the attacks of the musquitoes.   Three
nights successively I walked the beach, the moon
being nearly at the full : although I was little
disposed to do so, I could not help admiring
the delightful calmness around.   I treated Captain Mitchell in the usual way, but the calomel
had no effect.   About the fifth day of being
ashore, he felt anxious to be conveyed on board
again, and was at times quite insane.   When I
found that the calomel took no effect upon the
system, I felt assured he would not recover ; for
in several cases where it had failed, death had
most surely taken place in a very few days.   As
we were conveying him on board, a brig and two
schooners were seen off the bar with American
colours flying : the Captain of the brig afterwards came on board, and stated he was bound
here for palm-oil.   Captain Mitchell lingered

until the 10th, when he died, leaving our party reduced to three white men, two boys, and four negroes. He was interred on the west bank of the river, and a small board was placed over his grave by the mate, Mr. Robb. He was an excellent seaman, and took charge of the ship at Cape Coast Castle. Mr. Robb, the chief mate, succeeded in command of the vessel until the arrival of the Quorra.

The Americans had equally felt the excitement and interest consequent on the discovery of the termination of the Niger by Lander, and saw that the vast tract of country which offered a market for British goods might also be available to them. A few enterprising merchants of Rhode Island had therefore fitted out the vessels just arrived, consisting of a brig, named the Agenoria, of about two hundred and thirty tons burthen, and two schooners, with sixteen white men and several Kroomen. Their object, it appeared, was to trade for oil in the Eboe country with the small schooners, and to load the brig with it. The supercargo was a young man, whose unfortunate and dreadful fate will be mentioned hereafter. I was informed, that the means taken to induce men to join the American expedition were highly discreditable to the parties who engaged them. A great objection is felt by American seamen to the coast of Africa. I was assured by both men and officers, that when they were engaged, it was under the impression that the vessels were for the whale-fishery, and that the casks put up in shakes were for the purpose of containing the blubber oil; and it was not until they were actually on the coast of Africa that they knew their destination. The captain and supercargo had a copy of Lander's chart on a large scale, and were very sanguine of getting up the Niger into the Eboe country with the brig; a thing utterly impossible; for, in the whole distance of the first twenty miles up, there is in some parts only one fathom and a half water. However, the captain felt convinced he could go up the river, and proceeded up for the purpose of examining it in his whale-boat. He had not gone far, before he was satisfied that to take the brig up was impossible.

The supercargo, who had been on some part of the coast before, went to Brass Town to King Jacket. The king promised him protection, and returned to the ship along with him; and having obtained a great many goods from the Americans

on credit, Jacket separated from him to return.
It was intended that the small schooner should
trade in the Eboe country, and she had gone up
to Brass to proceed from thence by the creeks.
The supercargo left the Nun in a canoe, with
some of King Jacket's and King Boy's pullaboys,
in order to join the schooner at Brass; but when
he was near Fish Town, an accident happened
which not only terminated fatally, but completely
put a stop to the proceedings of the expedition.
The account given of this man's actions by the natives is as follows :—The supercargo, who had a
trade-gun with him, landed to shoot a small bird ;
when being disappointed at getting near it, he
returned to the canoe and laid himself down.
The gun unfortunately had been so placed, that
one of the pullaboys accidentally treading on the
lock, caused it to go off, when it burst and killed
the supercargo and two negroes. The subse-
quent conduct of the natives in the canoe ren-
ders their account worthy of little credit ; for they
plundered the mutilated corpse of the watch and
other things, and conveyed the former to an
officer then on board the schooner at Brass
Town, who was waiting the arrival of the supercargo. In addition to this calamity, King Jacket

would not allow this officer to leave Brass in order to communicate the melancholy intelligence to his captain until he was intimidated by threats, a week after the interment. It appears to me more than probable that an attempt was made to kill the supercargo, to prevent his going into the Eboe country, and thereby taking away a certain trade, and an old-established one, from their town, and in the attempt the gun exploded,—a very common occurrence among the negroes, as they charge the guns half full, in order, as they observe, " to make gun speak too much."

The captain of the American brig received the intelligence of the supercargo's death when performing the last ceremony over the corpse of his carpenter; and there being no medical assistance on board, he lost nearly all his hands. Fever had appeared among them, and in less than six weeks his crew of sixteen was reduced to five! To add to the misfortunes of these Americans, on one dark and stormy night, eight or ten Kroomen ran away with the brig's boat, with buckets and five boarding-pikes. A few months afterwards I engaged some Kroomen at Fernando Po, when I found that two of them gave out that they had belonged to the brig, but were badly used and badly fed, and therefore would not stay in her. This, however, may or may not be true; I merely give it as told to me. They sold the boat for some yams and provisions, in a creek near to Bonny; but it is fair to say, that these men conducted themselves much to my satisfaction during their long stay with me on the Niger, with no other Englishman on board than myself, and under circumstances the most dangerous and trying. A short time after the foregoing transaction, the captain of the Agenoria (Captain Pearce) died, and the mate took charge of the vessel and sailed for Rhode Island. Their whole expedition was thus unfortunately terminated : goods were given for nearly one hundred puncheons of palm-oil, not one quarter of which were paid.

Allusion has been already made to the custom prevalent at Brass, Calebar, and most parts of this coast, on the death of a king, chief, or any great man, of sacrificing the lives of two or three of his head wives. On going ashore a short time ago, I heard some mournful cries in the bush, and on inquiry from one of our Kroomen, found that a woman was going to be flogged.

Anxious to know the cause for this flogging, I
approached the place from whence the cries proceeded, which was about twenty yards from the
water's edge, on the sand, and I there saw a
woman lying chained to a log of wood, with her
arms and legs pinioned, awaiting the period of
high-water, to be launched into the sea, there
to become the unhappy prey of voracious sharks.
On inquiring what the poor creature had done
to merit such a punishment, I found that she
was one of the wives of a chief who had died a
few days before, and the brother had selected
her to suffer for having wished his deceased brother's death! I remonstrated with the brother
on the absurdity of the charge, and determined
to try to save the poor woman. I hastened to
Dido, the pilot, who also interceded; but the
chief made the following reply: " White man's
fash, no be black man's fash;" which being in-
terpreted, meant that they allowed us our customs, and we ought to allow them theirs. After
a palaver which lasted nearly three hours, owing
to the influence of Dido, who is a good man, the
chief agreed to liberate his sister-in-law, provided he might sell her to a slaver. We proceeded to the water's edge, which was now within

a few feet of the poor creature. Her arms and legs were disengaged; but, owing to the severe pressure which had been used, she could not walk until she was actually under her own roof. Indeed, so unexpected and unlooked-for had been her rescue, that she could not believe we were in earnest. She then pressed my hand with gratitude, and looked what her heart felt. In a short time after this, she completely recovered her spirits, and went on board a slaver, without any emotion, for she had no one to live for or no one to care for.

About this time a Spanish brig came into the river, to take in a cargo of slaves. She had been at Bonny a short time, and had been blockaded by his Majesty's ship Charybdis, Lieutenant Crawford. She got clear of the river quite light, having purchased her slaves, and sent them back from Bonny through the creeks to Brass Town. On the arrival of the brig in the Nun, the captain proceeded to Brass, and in two days returned with three hundred and twenty-nine slaves. They were shipped on board at four o'clock in the morning from King Jacket's canoes. She made sail at twelve, and was soon out of sight. In justice to the captain of this slaver, I must state that we met with more kindness and humanity from him and his chief mate than we did from the captain of the American brig. Being very short of provisions, two or three applications were made to Captain Pearce, who refused to let us have any, although he had abundance on board. Seeing no other alternative, and the mate and myself being left on board with only two more white men, I went to the Spanish brig and stated our want of provisions, and our wish to purchase beef. The mate, in a very polite manner, regretted we had not mentioned it before the captain had left for Brass; but gave us some wine and cordials immediately, and promised us bread. On the arrival of the captain with the slaves, at four o'clock in the morning, from Brass, an immediate attention was given to our request. At six o'clock, just at daybreak, the Spaniard's boat came alongside, with two bags of white bread, two casks of salt beef, six gallons of wine, and several other small articles, all equally acceptable; and when payment was offered, the generous Spaniard declined it, saying he should feel handsomely remunerated by my visiting two or three of his men who were then sick on board, and by giving him a little quinine, which I immediately did. He had lost thirty men, fifteen very recently, and knew full well the value of sulphate of quinine, a most important medicine on the coast. Most of his slaves were natives of Eboe and Nuffie. The brig was about one hundred and ninety tons, built very sharp and long, with two long six-pounders. The captain of this brig informed me he had made nine trips to Brass River alone, and this was his tenth, and that he had never been taken! He had been pursued twice by his Majesty's cruisers, and was once nearly taken by the Black Joke, the terror of the Spanish captains and supercargoes, but each time had made his escape by his vessel's superior sailing qualities. He was very nearly taken this time; for the Curlew's boats came in five days after he sailed.

It was now nearly the middle of April, and we had no accounts of the steamers. We were becoming heartily tired of our mode of life, having been upwards of five months at our anchorage without any intelligence of them; and, as may be supposed, time hung heavily on our hands.

About the 10th of April, the Curlew brig-of-war, Commander Trotter, was off the bar. This officer came on board the Columbine and kindly

supplied us with provisions, taking our despatches with him to forward to England.

About the 19th of April, we espied a canoe approaching us, on her way down the river, with a
union jack flying; and in a short time she landed
on the opposite bank, abreast of the brig. Two
or three individuals jumped out of the canoe,
when, by the aid of the glass, I thought that I
could distinguish Mr. Lander. We despatched
a boat to the party; and our joy may be easier
conceived than described, on finding we had not
been deceived, and that it really was Mr. Lander. We had, however, mistaken him for a mulatto—and with good reason, for he was much
sunburnt. Anxious inquiries were quickly made
after our old companions; but what was our dismay!—on asking how was Dr. Briggs? the reply
was, "Dead!"—how was Mr. Laird? "Very ill,
and not expected to recover!" — how was Captain Miller? "Dead!"—Mr. Jordan? "Dead!"
In short, officers and men were almost all dead.
Such intelligence was enough to shake the
strongest nerves: only eight or nine were left
alive, and four of these have since died.

Mr. Lander informed me that most of the
officers and men had been dead three months,

and that Captain Hill was in the canoe very
ill, and that he himself was also ill of dysentery.   I learnt from him that Mr. Laird had
been very ill, and had lost the use of his limbs:
he had gone to Fundah for a change of scene,
having lost his friend and companion Dr. Briggs,
who had died of dysentery in February.   Mr.
Lander also informed me that he did not think
Mr. Laird would return alive from Fundah.

Mr. Lander had come down the river for goods,
and, in the kindness of his heart, had risked his
own life to keep Capt. Hill company.   After a few
days he felt himself sufficiently recovered to venture in the long-boat to Fernando Po, where he
was hospitably received and kindly treated by
Colonel Nicolls, the governor.   After remaining
there a short time, and feeling himself much better in health, he went on board the Curlew brig-
of-war, Commander Trotter, who paid him every
attention, and cruised about a short time to
establish his health.   In a fortnight the Curlew
put into Prince's Island, where the Dove, American
schooner, was lying; Captain Pearce of the Agenoria was also on board of her, and as he intended going to the Nun on the following morning,
Mr. Lander took his leave of Captain Trotter,

and joined the Dove, when unfortunately, this vessel being light, the violent motion produced a relapse of dysentery. On her arrival in the Nun, on the 4th of June, I had then become convalescent from an attack of dysentery which I had contracted from my constant attendance upon Mr. Lander before he went to Fernando Po. When the schooner came to anchor, he sent for me on board, and I was sorry to find him very ill— much worse even than at his departure. He had brought with him a Mr. Dean to join the Alburkah, one European seaman, and Mr. Brown, a native of Cape Coast Castle, as clerk.

On Mr. Lander's arrival, preparations were made to go up the river in the Columbine's longboat and a canoe, as soon as his health would permit. He expressed himself to me in the most grateful manner for the kindness and attention he met with from Colonel Nicolls, at Fernando Po: the worthy colonel, he said, had administered to his recovery in no small degree by his kind attention and directions.

Captain Hill, who was very ill with a general decay of the constitution, formed one more victim to the terrible effects of the climate. I rendered him every assistance in my power; but from the first of his appearance, I entertained no hope of his ultimate recovery. He was twice tapped, and I was informed that he died on his way to Fernando Po, three or four days after my departure with Mr. Lander for the interior.

On the 7th of June, we were busy preparing to depart up the river. On the following day Mr. Lander's disorder not being quite so distressing as for many days before, I had hopes that he would soon recover.

CHAPTER III.

On the 8th of June, Mr. Lander finding himself much better, we determined on our departure ;
and I addressed a letter to the colonial surgeon
at Fernando Po respecting Captain Hill.   At
9. 30 A. M. we embarked in the Columbine's long-
boat, and the canoe in which Mr. Lander came
down the river, making a party of twenty-three
in number ;—two white men, Mr. Lander and
myself, and six Kroomen, being in the long-boat ;
and Mr. Brown and the remainder in the canoe.
The morning was delightfully fine, but very hot.
We left the Columbine under a salute of seven

guns, which we returned from our two swivels,
one being mounted in the bows, the other aft.
We had a delightful breeze up the Nun ; but
finding one or two sails had been left behind, the
canoe was despatched back to the brig for them.
At noon we were abreast of Eboe point. An aged
fisherman lives here in the midst of jungle on a
small elevated bank.   From him I purchased
some dried fish, giving him in exchange some
tobacco.   The river at this part is very narrow ;
on each side and in every direction are numerous
islands, upon which the pernicious mangrove-
tree (*khiz*, Lera Mangle) and the lofty palm (*cocos
butyracea*) flourish in luxuriance.   At 8. 30 P. M.
we made fast to a mangrove-tree for want of an
anchor.

On the following morning at 2 A. M. we again
got under weigh, after passing a most restless time
from the dreadful annoyance of musquitoes and
the excessive heat.   It being ebb tide, we found
ourselves drifted to some distance from whence
we started, by the rapidity of the current.

At 7 A. M. we were in sight of the first island, named Sunday Island.   This island is
situated in the centre of the river ; it is thickly
wooded, and appears to be at the extreme point

to which the tide reaches, distant from the Nun
sixty-four miles.   At the farthest extremity are
sandbanks, the first we have seen ; and it was the
first place at which we could disembark after
leaving the Nun, the country being nothing else
but swamps.   Mr. Lander appeared still to be
very ill.

About an hour after noon the river began to
widen, and we observed a gradual improvement
in the general appearance of the country :
sandbanks were now to be seen.   In the course
of the afternoon we had the wind south-west,
with very heavy rain; and to us it was very
cold, quite as cold, indeed, as a December
morning in England.   From the imperfect state
of the roof of the house which had been built
over the after part of the boat, I was compelled
to put over it my blanket-cloak, counterpane, &c.
to keep out the rain.   The depth of the water
close to the bank is seven fathoms : the grass
on the banks of the river is upwards of twelve
feet high.   At 5 P. M. we had heavy rain.   The
country appeared a little less wooded than lower
down.   The rain continued at intervals until 10
P. M.   At 7. 30 P. M. we stopped, and made fast
to the jungle.   This kind of anchorage is very

objectionable : independently of the risk of sleeping near such an unhealthy place, we were dreadfully annoyed by musquitoes. Owing to the heavy rains, everything under us was quite wet. Mr. Lander was still very ill, and could only take a little boiled rice.

At 12. 30 A. M. on Monday, June 10th, we proceeded by the light of the moon. About one o'clock we passed a small town, the first we have met with. This part of the river is very tortuous. At 6. 30 A. M. we came to another small town on the right bank ; and Lilly, a boy belonging to King Boy of Brass Town, was sent on shore to procure some mats to cover the house with. Several natives were assembled on the banks. Opposite this place is a creek, which leads to Brass ; and the Brass palm-oil traders pass through this creek on their way to and from the Eboe country. It was here, for the first time, that I observed an alligator asleep on the water.

About eleven o'clock we came abreast of a small town, named Eeckow, containing about eighty huts. At this town I landed, and was conducted by the chief to the Ju-ju, or palaver-house, where disputes are generally settled. He invited me to take some tomboh, or palm-wine, extracted from the palm-tree by puncturing. A large calabash, containing about six gallons, was brought, and its contents were served out to all present with great generosity. The huts are regularly built of a yellow clay, by the side of the river, and have a pretty appearance. I presented the chief a looking-glass, and he gave us in return two fowls. Rain coming on, I was glad to return to the boat, and we continued on our journey.

At about one o'clock, I was told by Mr. Lander that we were within a short distance of the Niger. A pleasant breeze now favoured us, and, gratified by our progress, we cheered our companions in the canoe, and continued under sail at about two knots per hour. It was not possible, however, to repress the thoughts of our actual position. We were then in a small boat, open to the attacks of the natives, and nearly four hundred miles from the steamers, in which many of our companions had fallen sacrifices to the climate. Such reflections were not very encouraging ; but our trust was in the Giver of all good for his divine protection.

At 7 P. M. we came to a town named Hooproomah. At this part, where the Benin branches off, the river is about half a mile wide, and sandbanks and huts are to be seen at a distance. At Hooproomah several canoes came off to us, bringing palm-wine : in one of them were three powerful-looking natives, wearing a piece of printed cotton round their loins. One of them impertinently asked why we did not go on shore ; and our interpreter was desired to tell them, that if they did not go away, we should fire into them. The swivels were immediately pointed at them, and they paddled away in great haste, somewhat alarmed.

On the 11th of June, at 12. 25 A. M. we again started on our way up the river ; but had some difficulty in inducing the Kroomen to work their paddles. A rainy and a dull foggy morning followed. We passed an inclosed place about three hundred yards wide in the river, for catching fish, nearly similar to those made on the coast of England for taking herrings. This morning we purchased a very large bunch of bananas (*musa sapientum*), which, with plantains (*musa paradisiaca*), constitute the principal food of the natives. Observed a great number of canoes at a distance. The river at this part is about twelve hundred yards wide, and presents a delightful appearance. On each side the banks are high, and the flourishing palm-tree is to be seen in great numbers.

At nine in the morning we came to anchor off a town named Oweecoodagra, the place where the steamers first obtained a supply of wood. Mr. Lander requested me to go on shore with a present of scarlet cloth, a looking-glass, and a bottle of rum, and to make arrangements with the king to have a supply of wood ready cut by the time the steamers came down. The king is named Hooquinah: he is an elderly man, with cataract in the left eye, and slight opacity of the right. The town consists of about forty huts, many much injured by the heavy rains. On landing, my interpreter conducted me to the palaver-house, where a low wooden stool was placed for me to sit on. A great number of the natives had of course collected to see me, and were squatted down round the house. In a few minutes the king made his appearance, dressed in a striped Guernsey frock, a silk hankerchief, and a piece of blue striped cotton, secured round his loins. I laid before him the presents which I had brought, and with which he seemed pleased, giving us in return a goat, a fowl, and a few mats to cover the boat with. When the natives found that we were in want of mats, several were brought to the boat for sale; but when the king heard that the natives were selling the mats to our people, he sent word to his subjects not to take anything for them; " For," said he, " if the white men, my friends, want mats, I will dash* them, not sell them." The morning being very hot, I rose to depart; when the king invited me to his house, and produced some excellent palm-wine, while two of his wives were peeping at us from a side apartment. At the back of the town were a great number of fine plantain and banana trees loaded to the ground with fruit.

A fresh breeze springing up, I took my leave of the king; and we departed immediately, and soon left the town out of sight. Mr. Lander still continued very unwell, the heavy rains we have had the last two days having in-creased his illness. At about half-past three in the afternoon we passed a small town named Hootwah, the natives of which hoisted a white flag and invited us on shore. There were seve-ral Brass canoes near this town, some having

* This is a common expression on the coast, and means " making a present."

puncheons in them belonging to the Bonny traders, who obtain them at Bonny from the captains of palm-oil ships.   Several other canoes were filled with jars, each holding about a gallon or two.— Soon after we passed another town, named Hooweecroodah : several large canoes were lying off it, laden with palm-oil.   Very heavy showers of rain fell in the course of the afternoon ; but the evening was fine, and we stopped abreast of a town named Anyammah.

The improvement in the appearance of the river and the country now became very evident. The banks were covered with the cocoa-nut tree, (*cocos nucifera*), plantain, banana, and there was also an abundance of palm-trees.

On the 12th of June, at a quarter past 3 A. M. we left our halting-place, the canoe taking the boat in tow.   It was a dull rainy morning, and the current was increasing against us.   At about half-past nine we came abreast of a town named Ahkibberee.*   The chief came alongside with a present of a large bunch of bananas, a fowl, some yams (*dioscorea bulbifera*), and plantains, and took in return for them a few beads, a bottle of rum, and two looking-glasses.   His name was

* Mr. Lander was shot near this town.

Osseeaber : he was desired to have wood ready
cut in a month's time.   His town consists of
about fifty huts.—At ten the morning was fine,
but excessively hot.   Passed a large sandbank on
our right, the reach of the river being nearly three
miles long.   At 3 P. M. another town was passed,
named after the last from some of the natives
having settled there.   In the afternoon we had
heavy showers of rain, and we passed another
town and village.   The river at this part is about
three quarters of a mile wide.   In the evening
we saw the heads of two hippopotami for the
first time, and at eight we stopped at a sandbank
for the night.

The following morning, at three, we proceeded until eight o'clock, when we halted on a
sandbank, to repair the house over the boat.
Several boxes of goods were taken out of the
boats to make room, and several articles completely wet for the purpose of being dried.   The
wet articles, as cottons, looking-glasses, and
clothes, were spread upon the sandbank ; and
muskets, pistols, cutlasses, &c. were strewed all
around.   We had been on the sandbank about
half an hour, in the full heat of the sun ; some of
the people were employed washing, some cooking,

others drying clothes, and the Kroomen were repairing the roof of the house,—in fact, the scene reminded me of a gipsy halt; when on a sudden several voices exclaimed " A war-canoe! a war-canoe!" and on looking towards the next town, named Bilbarrowkee, I observed a very large canoe, the natives in it being all armed. On approaching nearer, I ordered our men to take their arms; and they were in an instant drawn up to receive our visitors. Mr. Lander was unable to raise himself from his bed without assistance. When within thirty yards of the boat, taking Lilly the boy as interpreter, I advanced to meet them with a double-barrelled fowling-piece in my hand, and armed with pistols. Lilly hailed them, asked what they wanted, and desired them to go away. They replied, that they wished to see the white men. I desired him to tell them to go back and come in a small canoe unarmed, and we would speak to them. During this time some of our men were placing the goods, &c. back into the boat; while about fourteen or fifteen, doubtful how matters would turn out, were armed and prepared to defend us had an attack been made. After looking very cautiously around them, and seeing us so well and instantly prepared, the canoe departed to the town, and we shortly after resumed our progress. In passing Bilbarrowkee, we saw merely a few natives, although we fully expected every moment to be attacked.

Shortly after a canoe came off to us from Hypoteammah, a town situate on the left bank of the river. The king's name is Burneemeah, and he is brother to King Jacket of Brass. He sent us a goat, and we obtained twelve mats at two strings of blue cut beads each. Our house was now in a more habitable condition, and the muskets were neatly arranged on each side of Mr. Lander and myself.

At 3 P. M. we passed a creek leading to Bonny : on each side of it is a neat town, named Barrambee. At 8 P. M. we stopped, having had rain with heavy thunder and vivid lightning during the evening.

At 3 A. M. on the 14th of June, we continued on our journey in the wrong channel, and were drifted considerably by the strength of the current. At 10 A. M. we passed a town named Subercrebbee. When Mr. Lander passed this town a few weeks ago in a canoe with Captain Hill, the natives ran towards the canoe, and he was obliged to order it to be pushed off from the land. The natives were armed with muskets and spears.

We passed a few fishing canoes; on seeing us, the natives paddled on with great rapidity. Nearly opposite Subercrebbee is another branch leading to Benin, with a fine opening at its commencement. At this place the current was very strong, running at the rate of nearly four knots per hour. We were two hours in passing the entrance of this river, using sails and oars. We landed on a sandbank abreast of a town named Oppocoomah to breakfast, where the king's brother brought a goat to us as a *dash*. The natives of this town are well provided with muskets, spears, and knives, worn on the left side. One of the men brought an elephant's tooth for sale, weighing about forty pounds; but he wanted an exorbitant price for it, and it was not purchased. At four o'clock we proceeded forward, and on the right bank passed a town named Momotymiahmah, and several other small villages. This evening there was very heavy thunder and lightning, with rain: the current being very strong, we were unable to fetch the shore without drifting considerably. We made fast round some fishing stakes in the centre of the stream.

At 1 A. M. on the following day we resumed our progress, and at 6 A. M., from the effects of a violent current, to our mortification, we found ourselves below the place where we had passed the preceding night. We landed to breakfast on a sandbank, from which we saw several canoes loaded with palm-oil. At 3 P. M. we were abreast of Little Eboe, the natives of which town attacked the steamers going up, in consequence of which it was destroyed by them. It is now partly rebuilt, several of the huts standing quite back in the wood. The spot from where the greatest number of shots were fired by the natives is at the lower end of the town, and well protected by large trees. A great number of natives were assembled on the banks before the town. The town is situate on the right bank of the river, and consists of about thirty huts.—A little above this town is another, named Ahquimbra, abreast of which the steamers were supplied with wood. The king came alongside, and dashed us two goats and some plantains; some beads and rum were given to him in return. A tooth of a hippopotamus was purchased with a few beads. We lay to a short time near an island, but, owing to the musquitoes, were compelled to proceed.

On Sunday, June 16, the morning was fine, with a light breeze. We passed a town on the right bank which had a neat appearance, and halted on a fine sandbank nearly fifteen miles long. The current was very strong all day, and we stopped at night on a sandbank.

The morning of the 17th was dull and rainy. We halted on a sandbank opposite three towns, the huts of which are built of a yellow kind of clay, and have a very neat appearance: a great many of them had verandahs over their doors. A number of natives were assembled on the banks looking at us. The name of the largest town is Ingliammah. There are a greater number of natives at these towns than any I have yet seen. On our stopping, one of the chiefs dashed us some plantains and fowls; and we had a great number of natives round us. My fowling-piece was suspended from the roof of the house in the boat, and the chief perceiving the lock different from the muskets, being on percussion principle, asked permission to examine it. Having placed caps on the nipples, the chief and natives were quite startled at the report they made. I then charged it with powder; and when the natives heard the report, they set up a loud shout, saying "the percussion cap made gun speak too much." At this place we encountered a very strong current. Having passed several towns on the right bank of the river, we halted, as usual, on a sandbank for the night, which was showery.

The next day we halted near a town named Ohdonah, on a sandbank; the name of whose king was Egambia. After we had proceeded several miles this morning, we discovered that a mulatto boy had been left behind. The canoe was despatched in search of him, and in the mean time I loaded my fowling-piece, and shot a few sand-larks. When the canoe returned, we found the boy had been asleep on the bank, and had not heard our morning gun,—a signal we invariably adopted every morning to awake the people who slept ashore.

We passed some high land on this day, the banks being thirty feet high, the current strong. The width of the river was about three quarters of a mile; and the weather was rainy and foggy as we stopped near a town named Ohdoney,

which the current made it difficult for us to
reach. A native hailed our interpreter, and told
him, that if we would send some rum, cloth, and
powder for Ju-ju, Ju-ju would make fair wind!
This, as may be supposed, we declined doing;
and our interpreter pointing to our swivels, we
told him those were our Ju-jus. The fellow
did not like the allusion, and walked away. In
the evening we anchored, about seven, on a
sandbank.

During the night a very heavy dew fell, and
the morning of the 19th was foggy. I awoke
with a most excruciating pain in my limbs and
side, owing to lying in my wet clothes upon un-
even boxes. Jowdie was despatched for some
bamboo to make a kind of frame to support the
bed. Mr. Lander was much the same, but
stronger; and Mr. Dean, who accompanied us
to take charge of the Alburkah, was now very ill.
The river in some places is from twelve to fifteen hundred yards wide, the banks on each side
being covered with verdure and the richest foliage. The reaches of the river are ten or twelve
miles long, while here and there are to be seen
towns, with their brown-topped huts; some of
the natives were dressed in cottons and silks of

English manufacture, and many were in a state of nudity. About eight in the evening we anchored alongside a sandbank.

A great deal of rain fell during the night. About five o'clock, on the 20th of June, we again proceeded. Mr. Lander improving, and myself much better. About nine o'clock we were abreast of a large branch of the river, said to run to Benin. I believe this is the largest Benin branch which runs out of the Quorra. On the opposite side of the river we observed a multitude of black countenances, with white teeth, anxiously watching our progress, but apparently afraid of being seen. Our interpreter hailed them to bring off some mats, fowls, eggs, and yams, at the same time telling them not to be afraid. In a few minutes an amazing number of canoes were alongside of us, some having Indian corn (which is much cultivated here)—some with plantains, yams, fowls, eggs, goats, and mats, and red pepper. The river presented the appearance of a fair, and such a noise and confusion of tongues I never before heard. I counted upwards of seventy-eight canoes; several were very small, containing one and two men. The people sit astride their canoes, with their legs hanging in the water; a dangerous practice, from the risk of their being bitten by the alligators. The town extends into the bush a considerable distance along the banks of the river; it is named Owhyha, and appeared to contain about two thousand inhabitants. I observed a great number of interesting females paddling their canoes with great dexterity. The appearance of several of these sable beauties was particularly modest; more so, indeed, than any I have yet met with. In one canoe was a very stout woman with umbilical stroma, (rupture): her ancles were ornamented with immense rings of ivory, about five inches broad. She had a male slave in the canoe, a native of Kacundah, or Ibbodah, whom she repeatedly desired us to purchase, and was much disappointed when we gave her to understand that we never purchased human creatures. A gun, some powder, and a few yards of printed cottons, was the price demanded for this poor fellow.

Being much annoyed by the noise of so many people, each striving against the other to get alongside our boat, and frequently upsetting each other's canoes, I resorted to an infallible method of getting rid of them. I presented the muzzle of the aftermost swivel at them; on seeing which a general yell was the signal for departure—canoes were overturned, and a great number of the natives jumped overboard. One poor fellow, who had dived on the first alarm, ventured his head out of water, and seeing me still at the gun, uttered a cry, and again dived, and arose a considerable distance off. The natives paddled off with such rapidity, that it resembled a regatta of canoes. It was an amusing scene; but the plan effectually relieved us from further importunities, and in a few minutes the canoes were a mile and a half off.

This morning we observed two hippopotami coming down with the current. We fired canister shot at one, and it seemed to be wounded, as in a few moments we observed two swimming in-shore, and in the next minute we heard one blowing or bellowing something like a cow, near to the boat. We passed sandbanks ten or twelve miles in length. This morning we observed three canoes at a distance: when they saw us, they turned back;—we concluded they had come from Eboe to reconnoitre. At 11 P. M. we came to an anchor abreast of a sandbank. The weather fine, and a strong current: Mr. Dean sick, and also four blacks.

At 5 A. M. on the 21st of June, we again started, and found the current about three knots per hour. The morning was very fine, and we landed to breakfast on a sandbank, where we saw an immense number of birds of the swallow species, with a long beak. Their nests appeared to be in excavations made in the sand. A number of hawks were hovering around, apparently watching them. I placed my hand in one of the holes, and found a nest containing six young birds, partly fledged. As I was looking at them as they lay on the sand, about a dozen hawks darted down and carried them off. The sandbank resembled a rabbit-warren, the holes being so numerous. In the afternoon we saw four hippopotami on an island: after looking some time at us, they retreated very slowly into the water, and we observed them shortly after swimming to the bank of the river. The current being very strong, and our men very much fatigued, we were drifting down the river, when a canoe from the town of Otiyha, with a few Eboe natives in it, kindly assisted us: they supplied us with some Indian corn and a few cocoa-nuts,—a very acceptable present in our estimation.   In the course of the day we had heavy rain, with thunder and lightning.   Continuing on after dark, we passed three towns, one on the left bank, the others on the right bank of the river ; the names of the two latter were Etoaka and Adowah ; after which we stopped for the night.

We again resumed our tedious journey at five in the morning, which was then fine, but soon after we had heavy rain.   At 8 A. M. we halted on a sandbank, opposite a town named Amoorah, from whence we could descry some lofty trees near Eboe.   On this sandbank we found thousands of the same kind of birds that we saw the day before, and several hundred eggs lay quite uncovered.

A curious circumstance occurred on this day, which may serve to illustrate the difficulty of ascending the river in a canoe.   About 5 P. M. yesterday we were drifted several miles by the current, and crossed from one side of the river to the other.   Straw-Hat, one of our Kroomen, was tracking the boat along the bank, when he slipped into the water, and struck his foot against something hard, and immediately dived to find out what it was, and, to our surprise, brought up a basin which we had lost overboard before daybreak in the morning, nearly thirteen hours before.

Mr. Lander was much better of dysentery, but complained of debility : Mr. Dean was also better. The weather became fine about nine; but we had a severe hurricane, which lasted upwards of an hour. We were driven ashore by it, and came in contact with a stump of a tree concealed under water, which very nearly capsized our boat.

The river at this part is about two thousand yards wide. At 8 p. m. we halted for the night on a sandbank. The men had exerted themselves more than usual this afternoon, so as to have reached Eboe by the night; but owing to the strength of the current we were obliged to come to, distant from Eboe four miles.

This evening we passed a plantation, with very lofty trees on the side near the water. The plantation belongs, we are informed, to an opulent trader, no less than an Eboe gentleman. Yams, plantains, bananas, Indian corn, and pepper are grown in abundance on it. The Eboe people also rear bullocks of a small breed, goats and fowls, which are bartered in exchange for powder and cottons, with the natives of Bonny, Brass, Benin, and, I believe, Calebar. While we were at anchor the natives of Eboe kept up an incessant firing of muskets, which, our interpreter said, was in consequence of its being some Ju-ju day.

Our journey thus far had been very tedious and tiresome; the weather had been very unhealthy, constant rain with thick fogs; and feeling severely the confinement and exposure of a boat, we wished ourselves on board the steamers. Our black people became very indolent, and afraid of being attacked. We had now been absent fourteen days, but it was likely we should be as long again before we could reach the steamers. Owing to the quantity of trees on the banks, and the windings of the river, the winds carcely penetrates to its surface; and if it did, its serpentine form would render it fair one reach, and foul the next. The little wind we have had has been all round the compass; some days a refreshing breeze springs up by eight o'clock, at other times by eleven and twelve, but it soon dies away. Mr. Lander was now much better, and able to sit up: I was quite worn out with fatigue in waiting upon him night and day, as he would allow no person to cook or do anything for him but myself.

CHAPTER IV.

AT 10 A. M. on Sunday, June 23rd, we stopped abreast of a sandbank opposite to Eboe. As soon as we had arrived within a short distance of the town, we were met by a large canoe belonging to King Obie, containing twenty pullaboys on each side; which, with one steering at the bow, and another at the stern, made a total of forty-four men. They took us in tow, and soon placed us abreast of the town. The natives of Eboe are the most expert people with their paddles, and in the general management of their canoes, that I have yet met with: boys not more than four or five years of age are to be seen pulling with the greatest dexterity.

Mr. Lander and myself prepared to go on shore to visit King Obie, the same person who sold the Messrs. Lander to King Boy for goods to the amount of twenty puncheons of palm-oil. It commenced raining, and Mr. Lander therefore declined going, being also in my opinion too weak to stir. I then prepared to go alone with the following presents:—An arm-chair, lined with scarlet cloth, made by Hawthorne, the carpenter of the Columbine, a very ingenious man; a large looking-glass, two jars of rum, a pair of Turkish trousers of scarlet cloth and trimmed with dozens of bell buttons, and one piece of printed cotton. Taking with me four Kroomen, the presents were placed in the king's canoe, and after giving the king's slaves some rum, we departed for the shore. These slaves were fine youths, about nineteen or twenty years of age, with good flexible muscles, and such was their exertion that the perspiration was fairly flowing from them over the gunwale of the canoe. They kept time with their paddles by making a kind of hissing noise between their tongue and teeth, sounding like " Tchid, tchid, tchidda, tchidda."

The river is about one mile wide at Eboe, and we entered a creek about two miles long, at the extreme end of which we landed. The left bank has several steep landing-places, and between the Niger and the town of Eboe is a very large morass. A great number of natives were washing themselves with country soap, said to be made from palm-oil and alkali, obtained by the incineration of plants : it is of a darker colour than English soap, and of the consistency of soft soap. From the late heavy rains, the ground, which is of a dark and light yellow clay, was very slippery, and it was with difficulty we ascended the brow of the hill up to the town, the bank being upwards of thirty feet high. The distance from the landing-place to the king's residence is two miles. The morning was very sultry, and as I passed on, I was met by several persons who, my interpreter informed me, were the *gentlemen* of Eboe. It seems that they were proud to " crack fingers with (Eboe) white man ;" a custom they have obtained, perhaps, from our mode of salutation. They press the fingers, and in withdrawing them make a kind of cracking noise. I was met by so many, and all anxious to have a crack, that we were a considerable time before we reached Obie's residence, where we found a great number of men armed with guns and spears.

I was now conducted through a square and three court-yards, the doors of which, with the pillars, were ornamented with rude figures of men carved out of the wood. In the fourth and furthest court-yard, was a platform about three feet high, and covered with mats: at the back of the platform was a kind of throne, elevated about two feet, and covered with mats of a fancy colour ;—the walls of the yard were coloured yellow and red ;—the roof of the last court-yard projected out about twelve feet, and was supported by pillars of wood. In this inner yard, was what appeared to me to be a steam-chest, used by the king as a bath. The slaves and common people were ranged at the back part of the yard ; to the right were some chiefs and respectable-looking people, and to the left was a room occupied by Obie's head wife, a fine stout woman, so fat, she could scarcely walk, with anklets of ivory several pounds weight.

After waiting a short time, Obie entered from a door on the right, and in a most cordial manner shook my hand, and cracked fingers five or

six times. When we were seated, he embraced
me twice very affectionately according to the
custom of the country. King Obie * stands upwards of six feet high; he is very muscular, with
an intelligent countenance and dignified deport-
ment, and a quick and long step. He wore a pair
of scarlet trousers that fitted tightly round the
legs; no shoes or sandals; round his ankles were
seventeen strings of coral, secured on the inside
with a large brass button; he had a scarlet coat,
made by his own tailor, and cut like the coats of
the sixteenth century. On his head was a cap
made of red cloth, and trimmed with gold lace;
over this, a splendid general's cap, with the king's
arms on a beautiful plate in front, given to him by
Mr. Lander. The chair which we had brought was
placed in the centre of the yard; he examined it
very attentively, and then seating himself in it,
called for a looking-glass. He then proceeded to
examine himself in it, and burst out into a loud
laugh, thinking himself, no doubt, the happiest
of monarchs. He remained thus surveying and
laughing at himself alternately for some time, and
would have kept me longer had I not reminded

* The Eboe name for king.—Obesity is here considered
a mark of gentility.

him that his friend Mr. Lander was waiting, and that we were anxious to return. He then rose, took me by the hand, and having led me to his throne, began a conversation with his people, and required from me what he could do for us. My interpreter was desired to say, that in consequence of Mr. Lander being ill, we were very anxious to reach the steamers as early as possible; that our men were tired, and if he would allow us four pullaboys, he would render us a great service; and the presents which Mr. Lander had promised him should be sent back by the boys when the steamers returned.

After two or three minutes' consideration, he replied, " The boys no savy come back from that far place where steamers live;" to which I replied, they would return in a month. He said he would accompany me to the boat. He gave us two goats, some fowls, and eight or nine mats. He then invited me to take something to eat. Seeing a great number of people in the yard looking on, he ordered them all away, and in an instant not one was to be seen.

Shortly afterwards, Obie's two sons, fine youths, about eighteen and twenty years of age, entered the court, bringing in some stewed fowl and foofoo,* in clean wooden dishes, as clean as a dairymaid's in old England. I quite enjoyed it, although there was nearly a pint of palm-oil in it, and I had to eat it with my fingers. When I had finished, some water was given me to wash; and the young princes commenced eating the remainder. The eldest youth desired my interpreter to inquire what white men did with all the ivory they purchased. I showed him the handle of my sword, and pointed out a few of its uses. He appeared surprised, and said, "What a distance to come for ivory for such simple purposes!" and with much simplicity asked me if white men made houses with it.

Two slaves then entered with circular fans made of bullock's hide. They commenced fanning me, and I experienced a little relief from the excessive heat. Having sent word to the king that I was ready, he entered the yard dressed much the same, with the exception of a sheet which he now wore, and a small hat of crimson velvet trimmed with gold lace.

We left his house arm in arm, preceded by his

* Foofoo is pounded yam, similar to mashed potatoes, with palm-oil and Cayenne pepper (*capsicum annuum*).

drummer, who made a most horrible noise in his
avocation, and a youth bearing a horse's tail suspended to a pole.   On each side of us were a
great number of soldiers armed with spears and
muskets, and on the way to the water-side we
were met by upwards of five thousand of his
subjects.   It is the custom at Eboe, when the
king is out, and indeed in-doors as well, for the
principal people to kneel on the ground and
kiss it three times when he passes.   We were
met by several of the superior class of people,
who performed this submissive act without hesitation.   Among those who knelt was a female,
the largest woman I ever saw.   When this pon-
derous machine of human flesh knelt down, her
breasts touched the ground, and she even rivalled
the celebrated Daniel Lambert.   I could not help
stopping to look at this Hottentot Venus several
minutes.

The king's state canoe is a very large one, capable of containing eighty or ninety persons.

We now embarked.   A lofty pole bore the
union jack; and as we left the creek, the natives
crowded in great numbers to the different landing-
places to look at us.   King Obie appeared glad
to see Mr. Lander, and, as a matter of course,

begged almost everything he saw. Even when I was at his house, he wished me to give him a shirt, a pair of stockings, a pair of shoes, and suspenders; and "then," said he, " I shall be all the same as white man !" He also admired my scarlet tobe, and asked me for it; but it was convenient not to understand him. Mr. Lander presented him with a splendid officer's coat ; notwithstanding which, he was yet dissatisfied, and asked him for everything he saw. Another looking-glass was presented to him ; another was placed on one side of the boat, which when he observed, he asked permission to look at it, admired it, and handed it to his people to pass into the canoe with the greatest indifference ; and although he was told it was intended for the king of Attah, it was all the same to him. I was surprised to find him so covetous ; but it would have been bad policy to deny him anything after he had once seen it. A fine breeze springing up, we set sail, Obie promising to send the four canoe-boys after us.

The Eboes, like most savages, are a pilfering race. I witnessed one specimen of their ingenuity in this respect. During the time Obie was in the boat, some of his people were caught stealing a lamp and a pair of blanket trousers of

mine, which were found in the king's canoe.
When Obie was told of it, he pretended to be in
a violent passion, and ordered the canoe from
alongside.   When Mr. Lander and Captain
Hill last came down the river, two pistols were
stolen from the boat at Eboe.

It was nearly dark when we took our departure, and our canoemen made poor progress,
complaining of fatigue, three of them besides being sick; so that there was every prospect of our
having a long and tedious journey, the boat as
well as the canoe being very deep. Obie had not
sent the four pullaboys he promised; Mr. Lander was very ill, as well as two others; and under
these circumstances, I thought if additional assistance could be had, it was certainly desirable.
I immediately suggested my going to Obie for
the boys; which Mr. Lander approving, I hailed
the canoe and went on board her.

It was between nine and ten o'clock at night,
and a very heavy dew was falling as I proceeded
to land, and a most sickly mephitic smell issued
from the morass before the town. I walked up to
the king's residence, and found in the outer yards
two fires, round which were seated his slaves. I
sent word that I wanted to see the king, and he

soon came to me with a cloth thrown over his shoulders. He was very cordial, and said he would send the boys in the morning after us. I told him that would be of no manner of use, as we must have them then; and in a short time four boys were called in, who each in his turn knelt down before the king. He admonished them to be good boys, and work for white men until they died, and not to give over pulling strong. He then inquired what he was to call me. I told him " friend," and his attempts to pronounce the word as I spoke it were entertaining enough. He then adverted to the theft which was discovered at the boat, and informed me that he had made several inquiries, but could not detect the culprit; adding very emphatically if ever we caught them again, to let him know, and he would cut their heads off!

King Obie was particular in his inquiries after Mr. Lander, and looking at me earnestly, said, " You be ezogo (doctor) for my friend, and you pay him attention: when belly sick, bad palaver, very bad palaver :" I told him I hoped he would recover; and he added, " I no want you to stop now ; go to my freend and make him good again ; and when I see him again, I hope he will be quite good."

I reached the canoe between two and three o'clock in the morning very much fatigued, and after a pull of ten or twelve miles joined Mr. Lander in the boat.

Eboe being built some distance inland, near a creek which runs out of the Niger, is in the rainy season nearly surrounded by a morass. The Niger spreads quite above the bank, lining the west side of the creek. The huts are built in a straggling form, and the town contains upwards of six thousand inhabitants. They are powerful, well-formed men : some of them are of a light yellow colour, others very black, and their features highly characteristic of the negro· They are industrious in growing yams, immense quantities being sent down to the coast and up the river. They carry on an extensive trade in palm-oil and slaves : traders from Bonny, Benin, and Brass are constantly at Eboe purchasing palm-oil. The superior class of females wear immense anklets of ivory, seven or eight inches wide, and an inch thick: they are almost unable to walk with these immense weights round their legs. They are fond of coral or cornelian, red cloth, gums, knives, rum, and cowries. They wear their hair in various ways : some have it

plaited and twisted in a perpendicular form
above their heads.   The Eboe country is very
unhealthy, particularly to Europeans; and the
natives are subject to dysentery, diarrhœa, itch,
and a bad description of lepra and ulcers.   On
my going to and returning from Eboe at night,
the atmosphere was loaded with millions of
fire-flies, illuminating it as far as the eye could
reach.

On the following morning, we saw three hippopotami feeding about fifty yards from the water's
edge.   I had a distinct view of them : they were
as large as a cow, with very large heads, ears
like a mouse (very small for so large an animal),
skin dark, and heavy thick legs.   The morning
was sultry and showery.   I was happy to observe
that Mr. Lander was considerably better.   As
we pursued our journey, we found the width of
the river varied from one mile to a mile and a
quarter, and even a mile and a half.   At noon we
observed a little rising ground at Kirree, distant
about thirty miles from us.   In the morning we
had passed a very large island and sandbank, and
saw several hundred wild ducks.   In the afternoon a violent thunderstorm lasted some time.
We noticed a number of birds' nests suspended

from the boughs of trees out of the reach of monkeys and serpents: the birds were of the woodpecker species.

The next day we were in sight of Kirree. Some rocks were seen near the water's edge, their surface blackened by the effects of the atmosphere; they appeared to be of a sandstone formation. It was gratifying, after having been on the water so long, to be relieved from the monotony we had experienced, and to get a view of higher ground, and consequently of a more healthy country.

On the left bank of the river were several fields of corn, and the large-spreading tamarind-tree (*tamarindus Indica*) was seen.   As we passed on, we observed upwards of thirty small canoes with fishermen following each other with great regularity.   At this place we purchased several bunches of plantains for a button each : blue cut beads were also in great demand here.

In passing Kirree, the natives concealed themselves in the bush or hedges, and called out as loud as possible " Oh, Eboe! Oh, Eboe !" (white man, white man,) and invited us to go on shore ; but they were evidently afraid, as they kept themselves out of our view, and it was only occasionally that we saw their white teeth and jet countenances peeping from behind the trees. At this place we got into a wrong branch of the river while admiring a very pretty island: this branch is said to lead to Fundah. In a few minutes we were abreast of the identical spot where the Messrs. Lander were plundered on their way down the river. The sandbank is very large; and when we passed it, there were upwards of thirty canoes alongside, the traders in them being engaged in cooking.

On the previous morning we had a very narrow escape; and had it not been that we showed ourselves ready, and the natives observed our little swivels, the consequences would have been serious. In passing a large town on the left bank of the river, at the opening of a branch leading to Bonny, then dry, we observed two immense war-canoes: one was on our larboard quarter, and the other, at a distance of one hundred yards, came bearing down directly on us. Seeing that they were bent on mischief, I sprang forward to the swivel in the bows, and pointed it to the canoe on the larboard quarter. The natives observed me turn the gun round, and every one of them immediately jumped overboard, after throwing down their muskets and cutlasses, a great many remaining under water several seconds. In the mean time, those in the canoe ahead of us, as soon as they observed the natives of the other canoe jump overboard, steered their canoe more in-ashore. We continued on our course, and ordering our men to pull as fast as possible, were favoured by a breeze, which luckily enabled us to get out of their sight in a short time. The natives of this part of the country are great pirates, frequently plundering those of the small adjoining towns. It was impossible to pass this spot without reflecting on the exposure we were then subject to, and expecting that the same disaster which had befallen Lander, if not worse, might happen to us. The natives appear to view the white men's presence in their country with suspicion and distrust.

In the afternoon a canoe came alongside of us, with a brother of Abboka, King of Damuggoo, and an eunuch of the King of Attah. The latter was a fine-looking man of a most effeminate appearance ; he wore a small cap made in the shape of a lady's morning cap without strings, a striped tobe, with one of a darker colour underneath. They stated that when Abboka heard that Mr. Lander had departed to the sea-side, he was very

unhappy, and said he would never return again.
We learned from them that Mr. Laird had returned from Fundah ; and we were glad to hear
of his being alive, as Mr. Lander had no hopes
of his ultimate recovery. On the following day,
I found myself very unwell, and worse than the
day before, with a smart attack of fever.

On Monday, July the 1st, at five in the morning, we left our halting-place and passed some
beautiful lofty trees. The scenery at this place
was very fine, and we enjoyed the beautiful prospect it afforded us as we passed up the river. The
night was also fine ; and a little after midnight
we came to an anchor abreast of our usual halting-place, a sandbank. At 5 A. M. we were again
on our way, and about noon came in sight of
Adamey, which we passed about 4 P. M. although
the current was very strong. There are several
fine cotton-trees, three in particular (*bombax
Ceiba*), on the bank before this town. The huts
of the natives are situated back among the trees.
Soon after this we passed the spot where, as Mr.
Lander informed me, several of our officers and
men of the steamers were interred: the soil is a
red sand with a grey-coloured clay, the country
thickly wooded, and banks upwards of thirty feet

high. The crews of these vessels were as fine a set of men as ever left a port of England. Perhaps, if any error was committed, it was in selecting so many very young men; as I have observed that young seamen on the most trifling indisposition consider themselves dangerously ill, and give themselves up to despair, which invariably has a fatal tendency in this climate. As we passed the town, an immense number of natives were assembled on the banks, all anxious to have a look at the white men. We stopped at a sandbank, about two miles from the town, and gladly purchased some Indian corn ground and made into bread, called by the natives " cankey." We were informed that Abboka, the king, was at Attah.

Continuing our journey, on the following day about noon we halted abreast of a town for provisions, for which the natives asked an exorbitant price. The next day we despatched Mr. Brown ashore for provisions. The night was fine, with a brilliant moonlight. At 2 A. M. we continued on our journey. We had been informed that the Felatahs were at war with Cuttum-Curaffee, and had destroyed and burnt a town named Addacooda, and several others. I now found myself much better, having had a severe attack of fever : I was also happy to observe Mr. Lander was much better. The river the last few days has not varied much in width and appearance. We purchased a little salt to-day of a very black colour, having been subjected to a process of heat, by the natives, to prevent its deliquescence. The water the last few days has presented a very muddy appearance. We halted at 6 P. M.

A little after midnight we again proceeded. The weather in the morning was dull. At daylight we saw a canoe containing three persons, habited in tobes, looking as broad as diningtables ; and soon after, two beautiful cranes, a black one and a white one, flew over our heads. We passed several fields of corn millet (*sorghum vulgare*), in the course of the morning ; but it was not quite ripe. An immense number of birds were hovering over them. In some of the fields were platforms* about twelve feet high, on which boys were stationed under a thatched roof, pulling a string, to which several calabashes were attached within such a distance from each other that they would strike together by merely touch-

* This device is alluded to by Mr. Lander.

ing the string. These calabashes were filled with pebbles, in order more effectually to scare and terrify the marauders.

The morning was very fine, but oppressively hot. We purchased one hundred eggs, and out of that number only found fifteen good ones. In Africa the fowls are the property of the women. The natives do not eat the eggs themselves, it being considered criminal to do so, but allow them to remain in the nests to be hatched; which accounts for so many bad ones being met with.

As we pursued our course up the river before a fine breeze under our sail, our rapid progress appeared to attract the attention of some of the natives, who could not understand how it was the boat went so fast without the application of paddles. We repeatedly heard them exclaiming as we passed, " It is maghony, maghony (medicine, medicine)," not one suspecting that the sails were the cause of our progress.

Soon afterwards we were surprised on perceiving a canoe with the union jack flying; and on nearing it, we found it contained the brother of Abboka, King of Damuggoo, with two of his wives and about fifteen or twenty pullaboys.

This same man presented Mr. Lander a fine able-bodied slave, named Al Hadge, who had acted as pilot when the vessels came up.

About six in the evening we got into shoal water, and shortly after a violent tornado came on, accompanied by heavy rains, which soon satu-rated our beds. We remained on the shoal for some time; but, on getting off again on the fol-lowing morning, we soon reached an island before Attah. Mr. Lander was very unwell this morn-ing, owing to his having been wet last night. I myself felt very weak, and almost unable to stand a minute without falling. This I attributed in a great measure to want of exercise.

Attah is a very healthy-looking spot, situate on a high hill of sandstone formation. The huts are constructed in the figure of a cone. Several of the natives came down to the bank, some of them wearing ragged tobes, and others with pieces of common cloth thrown carelessly over their shoulders.

Last evening we passed the burial-place of the late Captain Miller of the brig Columbine, and the second engineer of the Quorra. The spot is a fine sandbank, on the right side of the river, a little below the town of Attah.

Soon after daybreak, Abboka sent us a present of a goat and some yams; to King Obie's boys he also made a present of a goat. In return, Mr. Lander presented him with a looking-glass, likewise one each to the King and Queen of Attah. Mr. Brown and Jowdie were sent with the presents.

We had anchored abreast of a sandbank covered with an immense number of huts built of dried grass and in a very wretched state of dilapidation. Upon inquiry, I found that the subjects of Abboka had occupied the bank while more substantial dwellings were erected for them at Iddah. It appeared that Abboka and his brother, the King of Iddah, after a separation of seven years, during the whole of which period there had been an uninterrupted war between them, had been reconciled by the exertions of Mr. Lander, who had brought about this desirable understanding upon quitting Damuggoo. Abboka had pitched his huts there, and now occupied the lower end of the town, which is named Abboka town. The houses of Attah are formed of red clay.

At 10 A. M. a fine breeze springing up, we departed, and made rapid progress under our sail until one, when we ran aground. A canoe was descried making in the direction of our boat, and as the messengers had not returned, we concluded they were in it. On coming alongside, however, Abboka and a messenger from the King of Attah appeared. Abboka is a fine old man, with a long grey beard, and is very stout. He was dressed in a striped tobe, and on his fingers he displayed six rings of silver and four of copper. He wore several charms of leather round his neck, containing seeds and scraps of the Koran written on paper by his Mallams: he also wore round his wrists four strings of Nufie beads. He stated that the king, his brother, had sent two horses to the water-side, and was much disappointed at our having departed without previously paying him a visit. He sent us a young bullock and a goat, to which the queen and the eldest princess had added a goat each.

Abboka left us about 4 P. M., at which time our canoe had not returned. Being extremely partial to rum, he had taken so much of it as to render it necessary for him to lie down, with his interpreter Al Hadge, and both went fast asleep. Shortly after we made sail, the canoe returned, having left Mr. Brown and Jowdie behind.

In passing under the hill on which the town of Attah is situated, a magnificent and imposing view of the Kong Mountains suddenly presents itself. There is something so grand in this prospect, that no language can do justice to a description of it. Attah, being so pleasantly situated, and possessing features so different from any I had hitherto been accustomed to see, struck me as being a particularly interesting place. The novelty of its appearance, and the magnificent prospects of the mountains, (the first I had seen since leaving my native land, many months ago,) excited sensations of pleasure and delight to which I had long been a stranger.

At sunset we halted for rest. At midnight we continued on our journey by the light of a bright moon, and when day dawned we were provoked to find ourselves close to the spot we had left five hours before, the indolent fellows in the canoe having been asleep instead of attending to the paddles. However, we made up for it by a run of about thirty miles.

On Tuesday, the following day, with the assistance of a smart breeze, we made good progress and were nearing the Kong Mountains rapidly. One mountain had a very remarkable appearance, resembling a sugar-loaf, covered with verdure and crowned with a dark black summit. The tops of the mountains are all of a tabular form (Kattam Katrassi).

On Wednesday, July the 10th, at about 10 A. M. we found ourselves abreast of the sand-bank where the Bocqua market is held; a little below which, on the right hand of the river, is the spot where Mr. Lander nearly lost his life. The reader may probably remember the lively description he gives of the danger he was in here. The arrow was aimed, the bow was drawn, and in another moment the fatal shaft would have done its business, had not the arm which directed it been arrested by a benevolent hand; and Mr. Lander and his brother would have been buried beneath a shower of arrows.

The branch up which we came this morning is about a mile in width. Just as the breeze was springing up, we had the misfortune to break our rudder, which gave us some trouble, and lost us some time. On our left was the craggy eminence named the Bird Rock by Lander. It wears a noble and commanding appearance, and is contiguous to the shore. We were admiring its fine bold form as we passed it, when our attention was suddenly attracted up the river, and, to our great astonishment and gratification, we beheld the two steamers under weigh approaching us, and shortly after they anchored within half a mile of us. When seen from a distance, the vessels bore a striking resemblance to barns, from their being thatched fore and aft, as a protection against the sun and rain; but the happiness we enjoyed on meeting them, in the relief that we now had from a confined boat exposed to the weather, besides the satisfaction of seeing our friends, may be better imagined than described.

CHAPTER V.

Rejoin the Alburkah. — Visit Mr. Laird. — His deplorable Condition. — Return of the Steamers to Attah. — Visit to the King of Attah. — Description of his Person and Palace. — Death of another Krooman. — The Steamers separate. — The Alburkah proceeds up the River. — Mount Caractacus. — Mallam Catab; his Character. — Straitened for Provisions. — Enter the Tchadda. — Lieutenant Allen taken ill. — The Alburkah runs aground.

A BOAT was quickly despatched to us from the Quorra, manned by Kroomen, with Harvey and Hector. We learnt from them that Mr. Laird had returned from the Tchadda in an ill state of health, and that since Mr. Lander had been down to the sea, four persons had died on board the vessels; namely, old Pascoe, Smith, a seaman, and Accrah, a Krooman, in the Alburkah; and on board the Quorra, Frying-pan, a Krooman, had died.

We now proceeded to the Alburkah, and were greeted by Lieutenant Allen on our arrival, who had been left in charge of her. In a short time

we went on board the Quorra, and I was shocked at the dreadful state in which I found Mr.
Laird: pale and emaciated to the last degree,
he appeared as if risen from the grave. He was
suffering from a disease named by the natives
" craw craw," — an inveterate form of scabies,
which, I am informed, is epidemic. In the vessels almost every white man and officer, and all
the Kroomen, had had it. I learnt from Mr.
Laird, that he had been up the Tchadda to Fundah, and had taken some goods with him, of
which the king, African-like, had fraudulently
deprived him.

Mr. Lander, Lieutenant Allen, and myself
having dined on board the Quorra, we returned
to the Alburkah in the evening, and as to myself,
I cannot say in very high spirits. We had
brought despatches with us for Mr. Laird,
and I was in great hopes of there being some
letters enclosed for me; but I was doomed to
bitter disappointment. Mr. Lander was also
very much disappointed at not receiving any.

On the following morning, I went on board
the Quorra to visit Mr. Laird, as I hoped with
proper treatment that in a few days I might be
of considerable service to him. A number of

canoes were at the market, and a Mallam
brought a fine elephant's tooth on board the
Alburkah for sale. Mr. Lander, who had been
considerably indisposed the last few days, was
now much better. The surrounding country is
hilly, the mountains being all of a tabular form
and very picturesque.

In the course of the day we visited the market,
which was numerously attended. Among the
crowd I saw several Mallams with clean blue-and-
white striped tobes. The commodities exposed
for sale principally consisted of Indian corn,
Nufie mats, elephants' teeth, tobes, spiced balls,
Indian corn, flour, horses of a small breed, slaves
in irons, blue beads of native manufacture, and
country beer: a few cocoa-nuts, goats, and dogs
were seen among the articles offered for sale.
The market is held every ten days.

On Sunday, July 15th, I had an attack of
intermittent fever. These attacks are becoming
less frequent, which I attribute to change of
climate. During the whole day I felt considerably indisposed.

Monday, July 16th, at 6 A. M. the Quorra
got under weigh for Attah; and shortly after, the
Alburkah followed her example. About half-

past eight, we passed a small town prettily situated on a rising hill, named Attacollico, off
which the steamers lay at anchor nine days
on their way up. The depth of water since
leaving Bocqua has varied from seven fathoms
to seven and three quarters mid-channel. At
11 A. M. we anchored nearly opposite the town of
Attah, and the Quorra also astern of us. I found
myself very unwell during the whole day, but in
the evening was somewhat better, and was able
to visit Mr. Laird.

The morning of the 17th was very dull. Late
at night, Soho, a trader from Cuttum-Curaffee,
came alongside with an elephant's tooth for sale:
he was obliged to come under cover of darkness,
as the King of Attah would allow no traders to
visit the vessels excepting his own. Soho said
he wanted to see Lander, and had run the risk of
coming thus far to see him, and sell his tusk to
greater advantage than selling it at Iccory market to the King of Attah's traders.

On the 18th, the weather in the morning being
extremely fine, we paid a visit to the King of
Attah. Horses were sent down to the water-
side for our use. These animals are of a small
breed, very high-spirited and swift. The saddles

are high, with a peak in front, and the bridles very complicated — a number of straps and ropes covered with charms being placed over and attached to the horse's head.

We had a very agreeable ride through Alburkah's town. In some places I observed grass rising to the height of eighteen or nineteen feet. We passed by the market-place, but I could perceive very few vendible commodities, except a small quantity of corn. The natives were armed with spears, bows and arrows; and several, I observed, wore a knife in addition. As we proceeded, we met a very interesting-looking woman of the Eboe country : she wore a pair of very large ivory anklets. On our right I perceived a gigantic tree of the monkey-bread-fruit-kind (*Adansonia digitata*).

After continuing our journey for a mile and a half or two miles, we arrived at the king's palace. To attempt to describe, or enumerate the houses, passages, through which we were conducted would be utterly impossible ; but I do not exaggerate when I say that their number amounted to upwards of fifty. I cannot pretend to say whether these passages, &c. led directly to the king's apartments or not, or whether we

were conducted by a circuitous route in order to impress us with an idea of the difficulties which strangers would experience should
they endeavour to penetrate into it without a
guide. Many of these passages were merely aper-
tures just wide enough to admit the body, and
frequent and severe were the thumps we received as we worked our way through them on our
hands and knees. At length, to our no small satisfaction, we were desired to sit down in a small
apartment, and a mat was prepared for our reception. Shortly after, the king's chief wife entered, accompanied by some slaves bringing in
some foofoo, and fowls dressed in palm-oil. Having first desired the intrepreter to taste the food,
we proceeded to satisfy the cravings of our own
appetites. We considered this precaution the
more necessary, as the King of Attah had manifested a hostile feeling towards Mr. Lander at
the time the steamers were aground at Addacoodah, and threatened to cut off his head if he
had an opportunity. He would not allow any
teeth to go on board, and forbade the natives to
furnish us with any kind of provisions. In addition to this, he despatched spies to the town of
Addacoodah, to watch all their motions. Pre-

vious to our quitting the vessels, we determined
that he should have more white men's heads than
one, if he was anxious for them : we all carried
our swords with us, and Lieutenant Allen and
myself each armed ourselves with a brace of
pistols.

After waiting a short time, we were conducted
through more apartments, and at length reached
the court-yard, where the king was ready to
receive us. This he did with a kind of stiff for-
mality and distrust, that determined us still more
to be on our guard against a surprise. He was
habited in a most peculiar manner. His dress
consisted of a rich silk tobe, resembling in colour
and pattern the skin of a leopard. Underneath
this, he wore two or three others. On his head
he had a large cocked-hat trimmed with tinsel
lace. His legs were adorned with brass fenders
projecting about four inches from the tibia, and
these leggings were finished with sandals covered
with red cloth. On his breast was a rude orna-
ment of brass, apparently intended for the sun ;
and a number of other brazen ornaments hung
suspended from his neck, with several dozen
strings of blue beads and alternate rings of coral.
In place of ear-rings, he wore plates of ivory,

which effectually concealed every part of his
features, leaving nothing visible but his eyes,
nose, and the centre of his mouth.   Two eunuchs knelt at his feet, whose shoulders served
him for a footstool.   On our right was an apartment, or seraglio, occupied by his women,
amounting to about two thousand: we saw
about five hundred of his wives, who were distinguished by wearing a quill in their hair, as a
symbol of their rank.   A trumpeter, stationed
near the house, at intervals of a few seconds
saluted us with a most discordant bray.

The morning being excessively hot and sultry,
we were soon desirous of returning; but the audience was not so easily broken off, and lasted
much longer than suited our inclinations.   Mr.
Allen, finding himself very unwell from his long
exposure to the sun, we mentioned the circumstance to the attendants, and were allowed to
enjoy the shade of the hut with the king and
queen.   I was seated near the latter august personage, who had stained her face, finger-nails,
and nails of her feet with henna,* to add to her

---

* Henna is a leaf, and is moistened and allowed to remain on the parts twelve hours. It produces a dark red
colour, and is also used to the face, giving it the appearance
of the polish of a bright mahogany table.

attractions. This is brought from the interior, where it is considered an indispensable requisite for the lady's toilet.

Mr. Lander informed the king before we left him, that a report had reached him, which attributed the death of Pascoe and others, at Addacoodah, to poison ; but his majesty denied all knowledge of it.* He ordered us a present of a fine bullock, and appeared very anxious for us to remain and establish a traffic. The conference having lasted two hours, we took our leave of his sable majesty. Abboka returned with us to the water-side, and we arrived safe on board, and, almost contrary to our expectations, with our heads on our shoulders.

This morning, a Krooman (Jack Savy) died on board the Quorra, being the third they had lost. I felt very anxious to make a *post mortem* examination, having been given to understand that the other Kroomen had died under very suspicious circumstances. They had been permitted to live on shore, and it was supposed that poison had been maliciously administered : Mr. Lander was decidedly of this opinion. The man

---

* Attah afterwards confessed to me that he had caused Pascoe to be poisoned.

in question had felt indisposed for some time: he complained of vomiting, a burning sensation at the pit of the stomach, thirst, debility, and occasional involuntary stools. His whole frame was very much swollen, particularly the legs and lower extremities. Having mentioned it to Mr. Laird, I went with the determination of examining the body. I found the Kroomen engaged in sewing up the corpse: I mentioned the motives of my visit, but met with a firm and decisive refusal. I expostulated, but they continued their work. I requested them to allow me to examine the face, and I would be satisfied; but this likewise they refused. Finding entreaties vain, I proceeded forcibly to remove the cloth from the face of the deceased. The body appeared very much discoloured, and presented a ghastly livid and appalling appearance, more so from being deeply pitted with the small-pox: a dark-coloured fluid was oozing from each angle of the mouth. I strongly suspect that the poor fellow died from ulceration of the stomach, having doubtless swallowed something of a poisonous and deleterious nature.

I now learnt that the Quorra was going to the sea-side, and not the Alburkah, as I had ex-

pected. From the 19th to the 26th of July, I
was much engaged in attendance on the sick of
both vessels, and in the agreeable task of writ-
ing to my friends in England. The despatches
were taken on board the Quorra, to be forwarded
from the sea-side.

The 27th of July, being the day appointed for
proceeding up the interior in the Alburkah with
Mr. Lander, I took leave of Mr. Laird, and at
12 o'clock we were under weigh. Mr. Laird had
rapidly improved during the last few days, and
though he still continued in a very weak and
debilitated condition, I was in hopes that the
sea-breeze would be of essential service to him.

The river was now evidently rising rapidly;
the sandbank on which we had halted on our
passage up, was completely covered with water,
as well as the huts upon it. Our friend Abboka met with a serious misfortune, the night before we left, in his house being set on fire, as it
was supposed, purposely. Since leaving Attah,
the depth has varied from one to two and
three-quarters fathoms. At 7. 30, P. M. we came
to an anchor off Attacollico. The following
day, being short of wood, we sent the Kroomen

ashore to cut some : the morning very fine with strong breezes.

On the 28th, at 6 A. M. we got under weigh, the weather being dull. At 10. 15, we came to an anchor abreast of the Bird Rock, and sent Kroomen to procure a supply of wood. A canoe came off, paddled by a native, who brought a few eggs and yams to Mr. Lander as a present. From the appearance of the rocks, the river had yet to rise sixteen or eighteen feet higher than we found it at this time, although the bank on which Bocqua market was held, when we were here a few days ago, was completely covered, and no traces of it visible.*

At 1 P. M. on the 29th of July, we got under weigh, and pursued our course up the river between the Kong Mountains, whose tops were crowned with dark luxuriant foliage, which added greatly to the richness of the scenery. On many of them evident traces of cultivation were observed, extending almost to their summits. On the western range was a majestic mountain, the highest in this part, to which Lieutenant Allen gave the name of Mount Ca-

* The rock on which Attah is situated is about three hundred feet above the surface of the river.

ractacus. When the boats first came up the river, an immense number of monkeys were seen on a spot a little below this part, which was immediately named Monkey Land. The Kong Mountains, and the rocks in this neighbourhood, are composed of black granite mixed with iron-stone. At about 6 P. M. we anchored off Ah-goojee.

At 2 P. M. on Tuesday the 30th, we were again on our way up the river, running much risk occasionally from rocks partly covered with water, and against which we often struck. In the course of the day we passed several lofty mountains on the west side of the river, of a tabular form, whose sides and summits wore a dark, gloomy aspect. We likewise passed a very extensive plain covered with rich verdure. Our general course was about north-north-east. At 6 P. M. we passed the spot where the Quorra lay so long at anchor; contiguous to which is a gentle eminence, the favourite haunt of the late Dr. Briggs. We came to an anchor off Adda-coodah between 6 and 7 P. M.

On the following day Mallam Catab came on board : it was strongly suspected that this fellow had poisoned poor old Pascoe. He is a fine, good-looking man, muscular and strongly built, and was respectably attired in new tobes. Some time ago, Abboo, a slave, was purchased from him, and his freedom granted to him, that he might act as interpreter, for whom he received fifteen thousand cowries. In the course of the day, a native with disease in the eyes came on board, and applied to me for relief: having made him a lotion, he presented me in return with some eggs and a fowl, which I could not refuse.

On the 1st of August we went ashore to the town of Addacoodah, accompanied by Lieutenant Allen, to take observations. We found the town in ruins ; it having been destroyed a short time previously by the Felatahs. It is prettily situated on and between large rocks of granite, which give it a novel and interesting appearance. As soon as we landed, we perceived a beautiful stream of water, clear and transparent as crystal, flowing from an aperture in the rock. Several houses were burnt to the ground, while in others nothing but the walls were left standing. Among the ruins of one I observed two fishing-nets ; and, while we were looking about, six of the natives appeared to observe us with some suspicion, and we therefore did not stay long.

While on shore, we saw a number of beautiful birds. Some of them had a plumage of a bright indigo colour; while that of others was of the richest purple, blended with hues of the deepest scarlet.

In the afternoon the Mallam's canoe came alongside. His people brought word that he had nearly blown his hand off, owing to the sudden ignition of some gunpowder which he had inadvertently placed near the fire to dry. He was in the act of turning it when the accident occurred. If this man had actually been the cause of Pascoe's death, he was justly served, and we should not have regretted the circumstance at all if his head had been carried away in place of his hand; but our suspicion of him was grounded on the information that he was employed by the King of Attah to remove those persons who are obnoxious to him. I sent him a lotion for his hand, which I considered was treating him with more kindness than he deserved

This afternoon we were destitute of everything eatable; and while regretting our hard lot, two canoes arrived, with goats, yams, corn, and country flour.

On the 2nd of August, before the sun was many fathoms high, as the Mallams term it, we were on our way to the Tchadda. The morning was beautifully clear and serene; the sun shone with uncommon splendour; and the small rocky islands in the midst of the stream near the confluence of the two great rivers, the Tchadda and Niger, gave an enlivening and cheerful character to the surrounding scenery.

We were now about to enter a stream which was (comparatively speaking) totally unknown, and where no white man had penetrated. What difficulties we might encounter before our return we could not foresee; and we trusted to the protecting care of Providence for a successful result to our efforts.

Our intention was, if practicable, to reach Lake Tchad from the Tchadda, which if we succeeded in doing, we should add a grand and important discovery to the results of our voyage. The natives considered it quite practicable, and Mallam Catab inquired if we were going to the Great Sea, meaning the Lake Tchad. To the westward of us was a long range of mountains.

Somewhat higher up, above the confluence of the
Tchadda, is the Kacundo Mountain, or Mount
William, beneath which the Alburkah lay at
anchor nearly six months.

A canoe was now seen to approach us very
rapidly, which, when it came up with us, we
found contained some Mallams of the King of
Addacooda. A boy was also with them, named
Friday, who, for some transgression or other,
had been punished on board the Quorra, whence
he had contrived to effect his escape, although
heavily ironed. Mr. Lander took him on board
the Alburkah, and made the king a present for
his attention to the fugitive.

At 7. 30, we entered the Tchadda. It appeared to be about half a mile wide at the confluence with the Niger. This width did not vary
much up to the spot where we anchored, about
ten miles distant from it. So far the river is
shallow, the soundings varying from one to three
fathoms. We passed several towns, situated on
the banks, the huts of which were of a conical
form, and very neatly constructed.

About 1 P. M. we anchored for wood. Adjacent
to our anchorage we saw several natives, stationed on the platforms erected to frighten away

the birds from the corn. By way of experiment, we allowed some of the steam to escape from the valve or blow-pipe boiler, on which they were so terrified that they leaped down from the stages and betook themselves to flight; and, for some time after, the towns were quite deserted. The current of the Tchadda at the confluence runs at the rate of from two to three knots an hour. I regretted to find now that Lieutenant Allen became very unwell. One seaman, and the mate Roose, likewise complained of indisposition.

The next day Lieutenant Allen, after an unpleasant night, in which we had suffered much from mosquitoes, continued very unwell. The morning was excessively wet; and the thermometer at 79°. Our progress had been diminished, from the engine not having sufficient power to stem the force of the current. We had now been a week in coming thus far from Attah. The distance from Attah to Bocqua is about thirty miles; and from the latter place hither, about thirty-five.

At 10. 50, we got under weigh: tried the rate of the current, and found it to be two knots per hour. A strong breeze sprung up from the northward, and continued a short time. The

reach we were now in lay about east three-quarters south; and with regret we saw that our
little vessel, at times, could scarcely stem the
rapidity and violence of the current. The river
is about fifteen hundred yards wide; the banks
are flat, and appeared low. The country near
the river appeared to be far inferior to that on
the banks of the Niger; although, during the
last few miles, the country around us had assumed a much more pleasing aspect.

The river is wider, and is deeper, during
the last few miles; in some places we had
four fathoms water, but in other places it appeared very shallow. This afternoon we disturbed a very fine alligator, the longest I have
yet seen, being about thirteen feet long. He
swam a very considerable distance before he attempted to sink. At 6 P. M. we unfortunately
ran into shoal water; soundings before had been
two fathoms, and there was nearly that depth of
water at the bows. Some unsuccessful attempts
were made to get the vessel off; and, night
coming on, we were obliged to desist until the
morning.

When on board the brig Columbine, Captain
Hill desired me to take charge of his property

in case of his death. I made an inventory of the principal articles, and copied them into a small journal, which I this morning delivered into the hands of Mr. R. Lander.

CHAPTER VI.

In the course of the morning of the 4th, the
Alburkah was hove off the shoal, and at 9 A. M.
got under weigh. In about half an hour we
passed a most delightful spot of ground, which I
named Frenchwood. Its elevation is about three
hundred and fifty feet; and it lies on the right,
or northern, bank of the Tchadda, and its extent
is about two miles. It would be a very eligible
situation for building upon, and seems well
adapted for the position of a fort. It appeared

428 VILLAGES OF OBWAH AND OBOFOH.

to be the most elevated spot of any on the bank
of the river for one hundred miles up. We soon
after passed a town named Obwah. A tornado
came on accompanied with heavy rain. At 11
A. M. we anchored abreast of a small village
named Obofoh.

This place is the residence of a man who had
fraudulently obtained goods to the amount of
seventy-five thousand cowries, by representing
himself as a chief of considerable importance.
At that time he resided on the banks of the Niger, in a town named Fundykee, and retreated
hither after his successful villany. Mr. Lander
despatched a message to the king, desiring him
to surrender the fugitive, adding, that in case of
refusal he would destroy the town. We were
informed that there were twenty elephant's teeth
for sale at Fundykee. While we were at prayers,
a canoe containing three females and several
men paddled alongside, bringing a present of a
goat and some Indian corn. In the evening we
showed a blue light, and sent up a rocket, which
produced so much panic among the natives, that
many of them precipitated themselves from the
canoes into the water, in order to effect their
escape. On this day Clarke, a man of colour

brought from Fernando Po, unfortunately fell down the hatchway, and fractured two of his ribs.

The 5th of August was a very sultry day, and our men were engaged in cutting wood. According to our expectations, the king of the village off which we were lying came on board. He was an elderly person, with a stout frame of body. Under two tobes, in which he was dressed, he wore another of striped silk, but of this only the sleeves were visible. A pair of Houssa trousers, the seat of which reached to the ground, covered the lower part of his person. He was accompanied by his son, a fine-looking young man, who was desirous to hear the " palaver " with the white men. After the usual compliments had passed on both sides, inquiries were made respecting the person who had defrauded Mr. Lander. The king answered, that he had gone to another village. Mr. Lander perceived the evasion, and in a peremptory tone demanded his surrender, threatening at the same time to destroy the town unless his request was immediately complied with. A canoe was then despatched for the chief of Fundykee, who shortly after made his appearance, attended by his slave, whom Mr. Laird had hired for six hundred cowries, with the intention of making him pilot; but the fellow absconded as soon as he had received the amount. Four of our men had been desired to mask themselves, to turn their coats inside out, and to make themselves as frightful as possible; and were directed to seize the fugitives, and handcuff them, the instant they came on board. Before the man could ascend the vessel's side, he, as well as his slave, the runaway pilot, were in irons. So complete and unexpected was the capture, that its effect was just what we desired; and to describe the fear depicted on the countenances of the natives would be impossible. The two culprits were violently agitated, and two of the natives in the canoe alongside threw themselves overboard, and swam a considerable distance before they could be induced to return. Had we been able to have preserved our gravity, I have no doubt that the example of these would have been followed by every native on board; however, sufficient was effected for our purpose. The chief immediately sent for what cowries he could command, amounting to twelve thousand; and his slave went away to fetch two elephant's teeth, which were delivered to us as a ransom for his master. It was, perhaps, fortunate for us that the minds of the people were so easily subdued. When the rocket was fired two evenings ago, the natives who were on the banks of the river, seeing it descend, threw themselves into the water, to prevent, as they supposed, their being burnt to death.

At 10. 30 A. M. on the 6th of August, we got under weigh. In crossing the river from Obofoh, to a village named Abasha, we sounded, and found four fathoms water. The morning was very dull and hazy. At 12 A. M. we passed the point of the island on which Obofoh is situate. The distance from this point to the town is about two miles. On the north side we sounded, and found two fathoms water. The average course by compass, as far as Yimmahah, is east by north. At this part the width of the river is about three quarters of a mile. We passed two withered trees, from which proceeded the chattering of five or six dozen monkeys, seated on their de-cayed branches. Passed a town named Eggoohla, and at 3 P. M. another, called Yimmahah, from which a road leads to Fundah, distant twenty-five or thirty miles.

At halfpast five P.M. we passed a small stream, the creek leading to Potingah, the port of Fundah, which is about ten miles from it by water, and the route by land from thence to Fundah is eighteen miles. An island of considerable extent lies at the mouth of the creek. Soundings were two fathoms and a half. The land we passed to-day was still low; and a long range, communicating with the Kong Mountains below the confluence, gradually disappeared into the plain, about thirty or thirty-five miles on the left bank, and about eight or twelve miles distant from the river. On the right bank there was very little elevated ground, and merely a few hillocks. The breadth of the river varied very little; the soundings being from five to seven and ten fathoms. At 5. 30, P. M. we anchored opposite a small town named Yoohai, and the only inhabitants we perceived were two or three of the natives passing through the trees. The distance we had run since morning we supposed to be sixteen miles.

As soon as we anchored, we despatched the canoe ashore for yams, with the interpreter. On their return they informed us, that the natives, concluding we were come to take their town, were in great alarm; that the women and children had been removed into the bush, and poisoned arrows were planted in pit-falls which they had dug to entrap us. The chief, however, came on board with a present; and, notwithstanding our ill forebodings, we had the satisfaction to find ourselves in possession of some good eggs.

Between 3 and 4 on the morning of the 7th, I was awakened by the noise of a canoe near the vessel. On its coming alongside, our old acquaintance De Lal, or Soho,* the same person who paid us such a late evening visit at Attah, came on board. His arrival was welcomed: he had been following our track from the Niger for three days, and had brought a bullock for us. He is a very shrewd and intelligent man, and stands six feet high.

About 4 A. M. we had a heavy squall, accompanied with rain. It was excessively cold; indeed, I may say, I never felt it colder in the month of December in England than I did at this time. In the evening I went with Mr. Allen on shore to take an observation; but feeling myself suddenly indisposed, I returned on board. There are several small villages within a short

* " Soho ' means an auctioneer or trader.

distance of each other, all of which are under
the dominion of the King of Jodah: his terri-
tories extend as far as Yimmahah.

At 11 A. M. on the 8th, we got under weigh,
and passed a small but very neat town on the
right bank of the river, at the rate of about five
knots and a half per hour; the force of the current against us being two knots.

We were now entering the Bassa, or No Bullock country, and passed an island, the name
of which I could not ascertain. The course of
the river from Yoohai is south-south-east. At
3 P. M. we passed a small town on the left bank
of the river, named Acatah, a little on the other
side of which is a small river, which flows within
ten miles of Fundah, bearing north-west from
where it falls into the Tchadda. Near it we
met with a sandbank which extends from the
town of Acatah a considerable distance up the
river. For the first time since leaving the Eboe
country, we observed two palm-trees growing by
the river-side. The banks were low and swampy,
and must render this part very unhealthy: the
appearance of the country near the sea-side is
very similar. In our progress we passed several towns situated at a small distance from the

river, at three of which the natives were armed
with bows, arrows, and knives. The arrows
were fixed in the bows ready for discharge. At
a town on the left bank, upwards of thirty
natives stationed themselves under a large tree,
presenting their arrows and brandishing their
knives in a menacing manner at us as we passed
them. We took no notice of them, and they
suffered us to pursue our course without moles-
tation.

This country is named Aquoto. We have
now come up the river sixty miles. Our course
this day was about east-south-east, the wind
south-south-east. We were detained about half
an hour by a large Fundah canoe, the owner of
which brought an elephant's tusk, weighing sixty
pounds, but demanded such an exorbitant price
for it, that we declined purchasing it.

The following morning was dull and cold, with
a strong breeze from the westward. Being very
unwell, I would not leave the vessel, and we
wished the interpreter, Al Hadge, to go to a
town about three miles distant for supplies; but
Abboca having been making war in this neighbourhood, he was afraid of venturing.

At 1. 30 on the 10th, we got under weigh;

course of the river, south-south-east. Passed an
island on the left; and at 4 P. M. passed another
island, lying to the right, covered with long
grass, and producing a few trees. The country
appeared high, and the river somewhat wider.
Our rate of sailing was four miles and a half per
hour ; strength of the current, two miles.

At 4. 35, we passed a beautiful spot of highly-
cultivated ground, laid out in rills. A number
of natives appeared on the bank, armed with
bows and arrows, the latter ready on the string.
In one part which we passed, the river was not
more than five hundred yards wide, being merely
a branch intersected with islands. The banks
are higher, and the course of the stream very
tortuous.

At 5. 45 P. M. we anchored off a town named
Ammagadah. A number of the inhabitants were
seen peeping at us between the trees. On letting go the anchor, the noise alarmed them,
and they fled in terror to the bush. The interpreter was sent on shore in the canoe. He found
the town deserted, and experienced great difficulty in persuading the natives to quit their re-
treats.

On the morning of the 11th, the King of Fundah's daughter came on board. She was accompanied by a daughter of the King of Addacoodah. The former is a most interesting-looking girl, with a very intelligent and expressive countenance. Her eyelids were covered with antimony; she wore coral studs in her ears, and was ornamented with Agary beads. Both the ladies reside at a village a little distance off.

A great deal of rice is grown on each bank of the river.

At 6. 30 A. M. on the following morning, I went ashore, taking with me my fowling-piece. I saw a great many monkeys of a greyish colour, but they were too wild to allow me to approach them within gun-shot. I shot a small but very beautiful bird; its feathers were brown and white, with silvery tips, brown iris, red beak, and four feathers in its tail eleven inches long. Mr. Lander styled it the " king of the wrens." We were in a fine plantation of Indian and Dower * corn, and saw in the distance a vast number of Guinea fowl.

At 12. 35, got under weigh. Course, east-south-east; soundings, from two to six fathoms.

---

* Dower, small red corn, so called by the natives, is the *sorghum dora* of Parsoon.

The tract of land lying on the right side of the
river is called the Domah country. At 3 P. M.
we passed an island about three miles in length.
At 4 P. M. passed two islands lying to the right,
bearing south-west. Just before reaching the
point of the first, we saw two very large alligators, and from this circumstance we named it
Alligator Island: to a recess formed by the point
of the smaller island we attached the name of
the Bay of Islands. At 5. 40 P. M. came to an
anchor in a part of the river which was not more
than three hundred yards in width. For the two
last hours we had been running between two low
islands, having by mistake left the main branch
of the river. In some places the width is only
sixty yards, and the Tchadda there bore a greater
resemblance to a creek than to a river.

The prospect around was now extremely wild
and dreary, although the country had been improving. No canoes or natives were visible, and
the only living creatures we saw were a few birds
and some baboons. The latter chattered, grin-
ned, and leaped from bough to bough in evident
terror and surprise. A suspicion that we should
be very much troubled with mosquitoes rendered
the thoughts of a night's rest in this place any-

thing but agreeable. We afterwards passed two small towns, but none of the natives were visible, though at night I could hear the sound of voices on the shore. Mr. Lander was very unwell during the last few days, and this evening I applied a blister to the lower part of the abdomen.

At 11 A. M. on the 13th of August, we got under weigh. A great deal of rain fell during the night, and the morning was dull and gloomy. Soundings, two fathoms. Mr. Lander found himself somewhat easier. When the Kroomen returned from cutting wood, Grey, their leader, informed us, that in attempting to go ashore at a small town a little below, the natives appeared on the banks, armed with bows and arrows, and resolutely opposed their landing. Accordingly they crossed the river, and landed on the opposite side; but, had our men been armed with guns, a trial of strength would, no doubt, have taken place, and the result might have been unfavourable to our proceedings.

We were informed that the King of the Bassa country had despatched a messenger to pay his respects to the white men; but the poor fellow, from some cause or other, dared not venture on board. He had been abreast the vessel several times, but always dropped astern after looking earnestly at her; and he finally departed, afraid to execute his orders. When he returned to his monarch, it is probable he gave some strange account of the white men and their vessel. We passed several palm-trees, and at 6 P. M. came to an anchor.

The river to-day has gradually widened as we have proceeded. At 4 o'clock we again entered the main branch; a circumstance which gave us no small satisfaction. All around us continued to wear a dismal and gloomy appearance. Since yesterday we had only had a glimpse of one small canoe. Our provisions had now failed, and we possessed neither rice, yams, nor plantains; neither was there a prospect of any town from whence we might obtain a supply. We had passed some palm-trees, and had seen an immense number of alligators swimming round us in every direction, one of which followed the vessel a considerable distance. The river has taken an ample sweep to the southward. In the afternoon we observed smoke rising from a spot situated a small distance from the banks, which gave us hopes that the next day we should be able to replenish our exhausted stores: but the thinness of the population, and the barren face of the country, led us to the conclusion, that the river is considered either unhealthy, or that little traffic goes forward here; otherwise we should have met with some natives, and seen more towns and villages. Mr. Lander found himself better.

On the next morning, the Kroomen went on shore to cut wood; but the natives again opposed their landing, and they were obliged to cross to the opposite bank for that purpose. The interpreter was also despatched on shore to procure a supply of provisions. On his return, he stated that we were contiguous to two towns, Dagboh and Obohbe, but at neither of them could he procure any provisions or information. He likewise informed us that the chief had fled on the first approach of the vessel. The chief of the smaller town was afraid to venture on board himself, but sent his daughter, a girl of ten or eleven years of age. She was entirely destitute of clothing, and wore nothing but a few strings of beads fastened round her loins. A small present was given to her, with which she returned on shore well pleased. The natives kept a fire burning all night, which we concluded they meant for their safety.

By information recently collected from the natives of Obohbe, it appears that we are two days' journey by land from Domah, behind the hills on the right bank, which is said to abound in ivory. The population of Domah is reported to be large, and the people industrious. We heard that the Sheik of Bornou had been at war with the King of Domah two or three months ago, and still possesses a hostile feeling towards him. It was thought that this would be a favourable opportunity for establishing a traffic, if the natives could be induced to bring down their ivory; and it was therefore arranged, that a messenger should be despatched to the King of Domah, to bring back with him three horses, should circumstances prove favourable, and to induce the natives to open a traffic for what ivory they were possessed of. Jowdie was accordingly despatched with this commission, and was expected to return in eight days.

Our anchorage was by no means a desirable one, the ground being low and swampy. To the north-east were hills about two hundred feet

high. We were still destitute of yams, rice, &c.
and with the view of seeing the village, in the
afternoon Mr. Allen, Mr. Dean, and myself went
ashore. A footpath from the water-side led us
towrads the village of Obohbe.

On our way we saw a number of girls occupied in grinding Dower corn,—or rather bruising
it to a powder between two stones, or slabs of
dried clay. The natives appeared very shy of
us, and evinced great timidity. We proceeded
a short distance along the pathway leading to
Domah, but quickly returned and followed that
leading to the town of Dagboh. To one whom
we met I presented my hand, but he shrunk from
it with horror.

On entering the town, I was much surprised
to find it of considerable size. The huts were
constructed of mud, in the shape of a cone, and,
instead of people, they appeared to be well
stocked with sheep, goats, and abundance of
poultry; indeed these were the only occupants,
for the inhabitants, men, women and children,
had all fled in dismay at our approach. We only
saw two of the natives in the town, and these
were armed with bows, arrows, and knives, and
wore the skins of the deer and monkey tied

around their loins like an apron. The natives whom we first met beckoned us to return, which we chose rather to comply with than to alarm them further by proceeding.

A great quantity of ochrus was growing around the town, but no rice was to be had here. The hills before mentioned are principally composed of ironstone, great quantities of which were lying in the foot paths. The natives generally are well armed with bows, arrows, and spears : the latter weapon is about seven feet long, and barbed at the end. They appeared to be the most savage, barbarous, and uncivilised people we had yet met with.—The evening, as we returned on board, was showery.

The following morning was also showery. There being as yet no sign of provisions, I volunteered to go ashore, with cowries, to endeavour to purchase some, accompanied by Mina, one of the Kroomen. As soon as I landed, to my infinite surprise, I beheld Jowdie, who by this time was supposed to be at Domah. He stated that he had proceeded some distance on his journey, when he was met by some Bornou soldiers, who opposed his progress, and, threatening to cut his head off if he proceeded, insisted on his return.

This was extremely mortifying. We were one hundred and four miles from the Niger, destitute of provisions, and with but little prospect of obtaining a supply, as the natives refused to furnish us with any. When we entered the towns, they deserted them, and retreated to the fields or mountains. Under these circumstances, there being no prospect of establishing a traffic, Mr. Lander deemed it advisable to return, much to my great disappointment, as I was in hopes of reaching Beeshle and Jaccoba, places of great trade, eight or ten days' journey up the river, abounding in salt called *trona* (a subcarbonate of soda).

We learnt that the fugitive chief of Dagboh had five elephants' teeth in his possession; but, notwithstanding our protestations of friendship, we could not induce him to trade with us. They imagined we were Felatahs, and consequently objects of suspicion and terror, as that people had carried on their predatory excursions up the Tchadda, and in their progress had set fire to several towns. We had now nothing on board but Dower corn, and having ground it as fine as possible, we had it made into bread, which, though somewhat coarse, was extremely sweet and palatable. At a small neighbouring village we could procure nothing but one fowl and some small yams, which cost us two hundred cowries each.

After taking shelter in a chief's house from a heavy shower of rain coming on, I went forward to Dagboh, taking with me my servant Lilly, Al Harr, and Mina. At the entrance of the town we found two natives on guard, one stationed on each side of the foot-path. On the left of it, they had placed a piece of iron about four feet long, to which was suspended four cowries, and some blue and white rags, as a fetish, to prevent my ingress. I advanced towards the men, who appeared by their manners to be surprised that the fetish did not prevent my advancing. They maintained their stations until I was within about a yard of them, when the one who stood by the fetish, casting a stern look at me, turned round, walked away, and was shortly after-wards followed by the other. I accosted them in Houssa, but they took no notice of what I said. Although the town contains more than two hundred dwellings, these were the only natives I could see. Abundance of goats, sheep, and fowls everywhere met my eye, and I could not but think it hard that we were to starve in the midst of plenty. After offering a reasonable price, we almost thought ourselves justified to take that by force which we could not procure by entreaties.

As we proceeded through the town, we could discern a straggling native here and there peeping at us; but on our advancing near them, they retreated to their yards and houses. The sun being very powerful, I went into the palaver-house, whither in a few minutes seven of the natives followed me, armed with spears, knives, bows and arrows. I was seated cross-legged, very composedly, when they entered; and after looking at me for a short time, they went away, and their place was supplied by others. Mina spoke to them, but they either could not or would not understand him. They put their hands in their ears, and by signs intimated their ignorance of what he was saying. After sitting a short time, one of the same men, acquiring an intuitive knowledge of the language, found that he could speak Houssa. We intimated our wishes to trade for ivory, and purchase provisions; and to make them more clearly understand us, we produced our cowries:

but all was quite inffectual. I repeatedly inquired for the king (*sulikeen*), but could obtain no answer.

We then left them, and returned to the village of Obohbe during a heavy shower of rain, where I found the natives in considerable alarm, in consequence of a dispute raised by the chief, because they had supplied us with a few eggs, and had held communication with us contrary to his direct prohibition; the Chief or King of Dagboh having threatened to give him *maghony* (poison) if he held any intercourse with us. The *malca*, or fourteen days' rain, had not yet set in; but in the evening we had two tornadoes blowing from the south-east, which was rather unusual: they were accompanied with heavy rain. Steam was ordered to be got up by six in the morning, and the vessel's course to be directed for the Niger and Rabba, having accomplished a voyage of one hundred and four miles up a river hitherto totally unknown except at its confluence.

It still remains to be determined whether or not the lake of Tchad empties itself into the Tchadda or Shary. If we are to believe the accounts of the natives, it does so; but it may be

that the Tchadda is formed by the rains from
the surrounding mountains. We could obtain
no information whatever which enabled us to
come to a just conclusion respecting it. When
full, the Tchadda must be a very fine and noble
river; but there are a number of islands, which,
with its shallow channels, here and there impede
its navigation.

On the morning of the 16th, at 5. 45, got
under weigh, and I was not sorry at the idea of
quitting this inhospitable country. Our fuel being green, it was some time before the steam was
up, which delayed us a little. We chose a reach
leading to the northward, and passed two towns,
but only saw one native : the poor creatures,
being terrified at our approach, generally concealed themselves.

About 8 A. M. we unfortunately got into shoal
water, with only one fathom round us, while we
were drawing five feet and a quarter. The weather showery and hazy. Mr. Lander, though still
indisposed, was better. We passed a tree the
blossoms of which resembled a honeysuckle ;
another bore a flower of the finest purple. On
the south-west side, the banks were clothed with
grass, about six or seven feet high. At noon we

were still aground, our efforts to get off the shoal being entirely unavailing.

At 3½ P. M. fortunately got clear of the shoal, and ran to the opposite shore, where we had deep water. At 3½ P. M. we anchored near a small town named Oruka, below which was a larger one called Corracu, situated on the north side of the river, both of which we had missed on our way up, in consequence of going up the other reach. Abreast of us, on the south side, were some palm-trees. The river at this part is about one thousand yards in breadth. A small island lay immediately above us. We observed several natives in the fields, two of whom afterwards came a few yards from the shore in a canoe, but almost immediately retreated. We desired the interpreter to invite them on board, but they proceeded to the island above us. Mina was sent ashore with a present to the king and chiefs, and on his return brought some yams and a goat, together with an invitation to visit the King of Corracu the following morning, when horses would be provided for our journey. This town was likely to afford us a supply of provisions, and relieved us from our anxieties on this score.

Within the last four months the Felatahs had plundered Corracu, carrying off goats, fowls, &c. together with a great number of the inhabitants, whom they take away with them as slaves. Hearing everywhere such accounts of the Felatahs, I had a great desire to see some of them. They are reported to be brave and warlike, and much superior to the generality of the natives; but their character is sullied by their cruelties, and by their neglect of their sick and wounded, who are unable to keep up with them on their journeys, and whom they leave to the mercy of their enemies, these never failing on their part to put them to a most cruel death.

Thermometer, 85°.

END OF THE FIRST VOLUME.